Rousseau and the Future of Freedom

This book examines Rousseau's conception of freedom and its significance for our modern technological world. Drawing on Rousseau's thought to explore the changing nature of authority, science and technology in modern society, the book's approach points to how Rousseau had a tragic conception of freedom, one that parallels the circumstances that characterize our own desire for freedom and democracy. Rousseau's critique of progress is integral to his thought in general and underrated when it comes to our own studies of science, technology and society. This volume refers to cases from the world of "free software" to consider our own predicament with how a flood of code and algorithms that is being wrapped around everything from our stuff to our food, to our bodies, our brains and – by extension – our freedom. As such, it will appeal to scholars of social and political theory, philosophy and ethics, particularly those with interests in science and technology studies and the implications of modern technology for freedom.

Eric Deibel is Lecturer of Science, Technology and Society at the engineering faculty of Bilkent University, Turkey, and is Assistant Professor at the political science faculty. He is a co-author of *Recoding Life: Information and the Biopolitical.*

Routledge Studies in Science, Technology and Society

Books in this series consider social science aspects of science studies. Authors discuss how science is socially situated and mediated, how science and technology are shaped by society and society by science and technology. Books will consider the social impact of new technologies.

Rousseau and the Future of Freedom
Science, Technology and the Nature of Authority

Eric Deibel

Routledge
Taylor & Francis Group
LONDON AND NEW YORK

First published 2023
by Routledge
2 Park Square, Milton Park, Abingdon, Oxon OX14 4RN

and by Routledge
605 Third Avenue, New York, NY 10158

*Routledge is an imprint of the Taylor & Francis Group,
an informa business*

British Library Cataloguing-in-Publication Data
A catalogue record for this book is available from the British Library

ISBN: 978-1-032-04523-8 (hbk)
ISBN: 978-1-032-04524-5 (pbk)
ISBN: 978-1-003-19360-9 (ebk)

DOI: 10.4324/9781003193609

For Product Safety Concerns and Information please contact
our EU representative: GPSR@taylorandfrancis.com
Taylor & Francis Verlag GmbH,
Kaufingerstraße 24,
80331 München, Germany.

Contents

vi   *Contents*

Acknowledgments

I've been told I take the acknowledgments in my books too seriously. This is the third time, almost a tradition, and one I'd like to continue.

While there is an immense amount of institutional support that goes into even a short book, I'll focus on the one that I consider indispensable to my work. There should be more institutions like the Nordic Summer University (NSU). The NSU organizes a wide variety of interdisciplinary symposia series every year and I contribute to one of them. The major topics of this book were first explored at its various events over many years. My work benefits greatly from regularly discussing ideas when they are still unpolished, when connections are tentative, and the goals keep shifting around. There are many workshops and conferences out there, but there are fewer and fewer places I know where this is possible. I count myself fortunate that there is, still, the NSU.

When all is said and done there would be no book and, maybe, anything else if it were not for Talya, my wife. This book was written under special circumstances. What was difficult, insurmountable at times, is now the past, because of her. She gave, carried me along, let me have my peace and even let me free when we reached the end. So I wrote a book about freedom, a freedom that I have because of her. This is selfish, of course, I am fully aware of this. Perhaps the book will find an audience, perhaps not. It does not matter, thanks to her I got to sweep aside the difficult circumstances of yesterday, and move on, to a life together, in the knowledge that she's there for me no matter what.

Introduction

Man, like software, is free but everywhere both are in chains.
  Forty years ago, the rhetoric of freedom found its way into the fields
of software development and coding. "Free as in freedom, not price",
so the saying went. And yet chains are clearly everywhere; a flood of
code and algorithms has wrapped around everything from your stuff
to your food, to your body, your brain and – by extension – your
freedom.
  A quick glance at the news is sufficient to show us that "freedom"
is at stake when dealing with software. Everywhere there are leaks,
breaches and hacks. It's about elections in more-or-less free societies,
about governments' surveillance of citizens' private lives, covert oper-
ations, fact-checking the enemies claims and it is about the massive
collections of data that are likely to include those little intimacies
that none of us would have chosen to share with others. Sometimes
it's about your health, with governments reaching far into our private
affairs, sharing data about our infections, injections and the digital
lives we lead. And yet, it is just another area wherein we are being
persuaded to give our consent; or, more realistically, to go along with
whatever is more convenient.
  That smart device most of us rely on, the one that is always close to
you, within reach, comes with many such conveniences. Most of these
simply require you to click and agree, and in exchange life continues.
Your health, privacy, security and, of course, speech are now all part
of the deal: a social contract of a sort, one that is less like a constitution and more like private contract, taking the guise of terms and
conditions. This is today's pact, even though clicking "yes" is a necessity, at least for most of us. Surely software, including social media,
is everywhere, as part of our lives and as integral to the deal between
societies and individuals. It is characteristic of our times even though,
tragically, it is at the same time nothing more than an interference, a

DOI: 10.4324/9781003193609-1

distraction to be gotten rid of, a moment in the way of getting on with what matters to us.

Clearly software is about freedom, it's nearly self-evident whenever we acknowledge that it is tied up to the new spaces of the internet and social media where public opinion and morality reveal themselves. It is increasingly where we show ourselves and join the citizenry, where the demos will increasingly take its shape – even if every single time we are subjected to the insidious drive to monetize whatever we do, think about and might become. In this sense freedom has long since moved beyond programming; it is no longer simply about the ones and zeros of coding, with its reach soon comprising nearly everything from the internet of things, self-driving cars, smart cities to autonomous missile systems and so on. Each of these subjects involves freedom and the same applies to how the languages of life are being programmed as the equivalent of text, with a meaning that can be stored, transmitted and shared as code. From your genes to your machines, food in your fridge, vaccines, pills, even bio-energy, in each case the same argument about freedom and software should apply. However, they do not; clearly each of these examples can show how difficult it is to refer to freedom as a meaningful concept in whatever context software has gotten a hold.

Tragic, certainly, and this is the principle reason to return to reading Rousseau, or rather to begin this book by paraphrasing the famous line that opens the text on the social contract. It applies directly to the predicament we're in; to how, in characteristically Rousseauian manner, we are being enchained.

Rousseau's tragic conception of freedom

Opening a page at random in any of the major political treatise by Rousseau will mostly likely show passages that are scathingly critical of "society" and how it's inevitable that we lose our natural freedom. In *Émile,* the educational fantasy that is simultaneously a comprehensive political treatise, the main character grows up to become a citizen by learning about conformity. Not even his famous "general will" is what it is often made out to be, an unrestrained endorsement of "the people", of popular sovereignty. And the same applies to Rousseau's relation to science and technology, his critique of progress.

Therefore, it should not be surprising to take Rousseau's tragic conception of freedom as our starting point. Those with even a passing familiarity with Rousseau's writing are likely not surprised, but such a reading does contrast with more casual references. After all, Rousseau is more popularly identified with optimism about human nature and idealism about democracy – as an advocate of popular sovereignty, a believer in human nature, the beginning of child-oriented creative education. While each of these areas are part of his writings, such characterizations mean little when considered without even a minimum of awareness about the intellectual context to which they belong, let alone how they fit within the wider context of his major works.

His published work began with a critique of progress, with "a discourse on the Sciences and Arts" from 1750 and is constantly discussed throughout his work. In other words, Rousseau's critique of modern society revolves around questions of science and technology; they were there at the start and are present throughout his work. This is, therefore, also our starting point; we begin with Rousseau's critique of progress and how it was written with a sense of urgency, or rather the sense of tragedy, that pervades his work. This is the general point of his first publication, the "first discourse" and it is reflected in the first line of the Social Contract ("he lives in chains") as well that of Emile ("in the hands of man, everything degenerates").

To some, perhaps such a sense of urgency and tragedy might appear as overstated when it is taken as a starting position for our own times. It is true that some of the major works are less polemic in style, notably the discourse on inequality and political economy. However, these concern themselves with topics contained within the other works, to be expanded upon. Accordingly this short book takes on a similar structure, beginning with the observation that a tragic conception of freedom is in order. It is not just that there are plenty of indications of crisis or that there is a need to commit to a meaningful conception of freedom. As mentioned, this is nearly self-evident. What is missing, however, is a sense that tragedy and a commitment to freedom are necessary, and logically, tied to one another, as Rousseau showed. Its implication, for us, is not that emergency and catastrophe should dictate the outcome, as a naturalistic fallacy wherein disorder turns into a call for order. Quite the opposite, tragedy is our starting point so that we commit to an approach to freedom that is neither gratuitous nor simply whatever it happens to be in practice.

The inspiration that we take from Rousseau is primarily methodological: affirming Rousseau's tragic conception of freedom, as our starting point, is a means whereby to reconsider to engage with the intricacies of today's analysis of the relation of knowledge and power. If we assume that a meaningful discussion of freedom is increasingly moving away from us, that our own experience of freedom resembles that of Rousseau, its implication is a need for a renewed commitment to freedom to accompany our fascination with ever more careful analysis of the inner worlds of science and technology.

There is nothing that is necessarily wrong with that fascination, it is a necessity and a moral imperative given how our relation to progress sits at the center of many of the crises and catastrophes that characterize modern life. At stake is not the fascination itself, but the recognition that freedom is at stake in our methods, in how we act on that fascination and how the understanding we gain of our predicament does not result in a meaningful conception of freedom. In other words, we find ourselves, once again, running towards our chains (in our analysis) and this will remain the case unless we recognize that a meaningful concept of freedom is at stake in our empirical commitments (as method).

Freedom and method

Maybe the search for a renewal of our own approach of freedom seems underwhelming to those who picked up this book out of a desire to find definitive ontological claims about Rousseau's concept of freedom. The same might apply to those who are not interested in our own predicament, reading Rousseau in all possible nuance as a precursor to later philosophers and within the confines of a specialist's approach to political theory. After all, this is how Rousseau's interpreters over the centuries have singled him out in discussions of the meaning of Enlightenment and these traditions continue today.

Accordingly there's no shortage of appreciation for Rousseau's critique, his counter-Enlightenment. In that capacity Rousseau became the principle liberator of feeling in the 19th century, the discover of intimacy (e.g., Schiller 2016, Arendt 1997), as well as the direct inspiration for rationalist philosophies, most notably those of Kant (Cassirer 1989), Hegel (Neuhauser 2003) and Marx (Althusser et al. 2007). Each of these philosophies has deep roots in Rousseau's critique of the doctrine of natural philosophy of his predecessors, as a precursor to their commitment to different types of critical reason. Accordingly, the meaning of Enlightenment is seen as a precursor to the age of "critical reason"; taking inspiration from Rousseau in various ways is essential for what followed, whether this is Kant's imperative, Hegel's dialectics, Marx' materialism or otherwise.

These readings continue today and this also applies to how Rousseau has been held responsible for the whole modern problem, for example when seeing him as emblematic for the "first crisis of modernity" (Strauss 1947) or personally held in contempt as the first modern man (Nietzsche 2003). This happened almost immediately, during his life time, when he was held complicit for the totalitarianism and violence, when the Jacobins put him on a pedestal and claimed him for their Reign of Terror (Burke 2009, de Maistre and Lebrun 1996). Also later, he would once more be treated in much the same manner. The Enlightenment revolves, in this sense, around the merger of rationalism and authentic experience, with Rousseau as the principle sentimentalist, a man of feeling, emotion and imagination that led to the horror of the 20th century.

Accordingly it's with Rousseau that this fusion appears, as fully modern for the first time, taking the shape of a synthesis between the traditions of Enlightenment and Romanticism in 19th-century philosophy. For example, Popper discusses the expansive role of the modern state by characterizing Rousseau as a dangerous romantic collectivist. Berlin sees Rousseau's appeal to self-realization as one of the primary inspirations for the many demands for control, indoctrination and conformity within the historical context of the re-organization of society, be it as a nation, according to class interest or as a type of democratic totalitarianism (see Berlin 2002, Popper 2002). Here the question, however, is not primarily whether Rousseau, as a key Enlightenment (or counter-Enlightenment) figure, should be positively or negatively reviewed, whether or not he should be praised as a precursor of the critical reason of 19th-century philosophical works. Our reading is that Rousseau is underappreciated as a methodologist of freedom when it comes to knowledge and power, which the subtitle of this book captures by referring to "science, technology and the nature of authority".

Who knew? Rousseau can help us re-orient ourselves towards a different type of critique of progress, one that is tragic but that at the same time renews our commitment to freedom in the company of the highly speculative futures that surround our devices, locking us into an ever firmer embrace, persuading us to opt-in to the hope that somehow freedom will come out of the desirable new features constantly marketed to us as gifts of progress. Of course, such a methodological reading of Rousseau implies that further discussion of "how" Rousseauian freedom should be understood; both transhistorically, in terms of how to position ourselves vis-à-vis Rousseau's historical interpreters, and as an inspiration for our own empirical commitments today. For now, however, our starting point is simply to point out that this setting includes the contemporary imperative of "science in society" and "technological culture", which is a reference to the constant concern over ethics, responsible innovation as well as the democratization of new technology.

Ostensibly freedom is part of this mix, and given its importance in public discourse, the suggestion would seem to be that political theory carries with it a great deal of legitimacy, rationally, romantically or otherwise. Yet, the opposite is the case, as can be observed quite easily when considering the field of science and technology studies (STS). This field will be considered indicative of a wider problem within the social sciences. To be specific, STS is indicative of how science and technology studied empirically in a variety of ways within the context of an intellectual framework wherein the relation to political theory is rarely appreciated. Rousseau is exemplary in this regard, as his critique of progress is integral to his tragic conception of freedom, while at the same time deservedly carrying the popular legacy as one of the inspirations of our own concerns over direct democracy and the role of the citizen. Advocating the latter implies, logically, a relation to the former, in the sense of strategically and selectively mobilizing the history of political theory, whether it is Rousseau or otherwise.

Therefore, our point is not that there is anything unique about STS today, that there is something that sets it apart from other fields wherein science and technology are being discussed. Rather, STS as an empirical field has been around for a while longer, which implies that it can be discussed as a sizable and relatively coherent field of engagement of the social sciences to science and technology. Usually this is a history that is traced to the 70s, but there are obviously numerous influences that predate its origins as a field, including older examples of science policy, citizen engagement and so on. What sets apart its origin, its current scale and its empirical focus is its highly peculiar relation to political theory. Specifically, the period of the scientific enlightenment and by extension modernity in contemporary empirical analysis can be examined in terms of its political theory by way of a well-documented and iconic debate over Thomas Hobbes. This debate has a unique (and peculiar) position within the curriculum of STS (see Shapin and Shaffer 1985, Latour 1993, 14–26, 1999: 263, 2015, 147–149, Haraway 1997, 23–27, Law 2008, 2015).

This status is well-deserved, given that Hobbes is part of an exceptional study by Shapin and Shaffer. This is a celebrated book, never absent from lists with the field's core literature. Indeed, there is no other figure from the early modern period of political theory, from before the 19th century, that has a status that comes close to that of Hobbes.

No one else is on the list, even if there are, of course, occasional references in various specialized histories of science. This might involve other major figures from political theory and some, Bruno Latour for example, might occasionally reminisce on the entire early modern period in general terms (see Latour 2017, 184–189). Nonetheless there's no other candidate than Hobbes to establish a concrete connection of STS canon to early modern political theory and, in that capacity, recognize what characterizes our own predicament when it comes to our own relationship to the scientific enlightenment and, by extension, our ability to critique our own claims on progress.

Perhaps STS is an inclusive field in many respects (see Jasanoff 2016), when it comes to contemporary issues like democracy, the public, the citizen and such topics. For this reason the abbreviation got adjusted, from science, technology studies to science technology and society. However, this inclusivity does not extend to political theory and specifically to Rousseau. It's exactly these concerns where Rousseau's legacy is most obvious, much more obvious than Hobbes in any case. Even more striking is the absence of such political theory within the setting of discussions of (bio-) constitutionalism (Jasanoff 2011), as the successor to earlier ideas about a natural contract (Serres 1995), a non-modern constitution (Latour 1993) or a technological constitution (Winner 1982, 109). Then again, it's not necessarily accidental. In the history of science there is nothing unusual about revisiting historical figures, paying special attention to their scientific contents and contexts, but this does not thereby imply a sustained relationship to political theory. Furthermore, the point is not restricted to an observation about historical scholarship, some type of lacuna in a specialist area. There are sustained concerns over STS and specifically its claim on empirical research. For this reason our focus on Rousseau, and his relation to Hobbes, allows for an analysis of how the latter became intricately tied to the controversy between the sociological approach of the "Strong program" and a Latourian re-interpretation that remains at the root of how his work has gotten traction across the social sciences and the humanities.

On the one hand, Latour's popularity today goes beyond STS and it is well-deserved given that science and technology are everywhere. As a consequence the debate over Hobbes and its extension to STS provides a unique vantage point, one that is not only applicable to STS but extends to the many other fields where Latour's ideas have been internalized. On the other hand, there are longstanding concerns over STS and its "stringent naturalistic monism" (Collin 2011), specifically the lack of awareness of the philosophical roots of STS, and, more generally, its disconnect from the major currents in political sociology (see Frickel and Moore 2006, Söderberg 2017). Within this context a focus on Rousseau does not simply extend how, in Latourian STS, it is Hobbes who invented "a Monist society in which knowledge and power are the same thing" (Latour 1993, 26). Indeed, this would imply a tragic conclusion, a counter-Enlightenment wherein we are left seeking to confront a comprehensive transformation of modern thought, one wherein a lack of freedom and democracy will appear to the modern subject as if it were its "second nature" (Horkheimer and Adorno 2002 [1944]; Horkheimer 2003).

Switching from Hobbes to Rousseau might easily be seen to imply that the latter's critique of progress is simply a precursor to any type of conception of science and technology result in a "one-dimensional" world (Marcuse 1991). Surely this is another way to collapse knowledge into power, but it runs into the problem that it is too convenient to reduce Hobbes and Rousseau to some type of categorical counter-Enlightenment. Latour's anti-Hobbesianism might make sense to the extent that Hobbes is irredeemable in his commitments to order, it is also overly convenient as a means whereby to create a sharp line between the past and our own attempts at Enlightenment. The same applies to critical theory, which should be recognized for being heavily invested in our self-understanding as subjects, a subjectivity that is not somehow defined in contrast to science and technology but comes out of their respective critique of progress and its relation to enlightenment (see Feenberg 1999). Our topic, however, is Rousseau, to whom the same applies. Even if Hobbes is not fully redeemable, in this sense, certainly Rousseau's major works can quite straightforwardly demonstrate that his critique of progress revolves around a relationship of the sciences to nature that is multiple, diverse, imaginative and carefully presented as a step-by-step process that is integral to his statements on the sovereign, the social contract, the citizen and all the rest of the conventional categories of political theory.

What more could we want for our own attempts at (counter-) Enlightenment? Our interest in a meaningful concept of freedom, for our own sake, is therefore by no means an attempt to establish some type of definitive origin story about the "real" Enlightenment. This is not what reading political theory is about, and with seriously reading Rousseau comes an orientation towards giving meaning to freedom that will carry legitimacy under tragic conditions, in crisis. It is inescapable today, we have to intellectually relate to the tragedy, emergency and loss that surrounds us, that is coming, as characteristic of our times. Therefore, this book seeks to respond to this predicament in terms of a call for a renewal of our relation to critique and to method. Our starting point, in this sense, informs our critique, one that affirms that there are many types of foundations and origin stories, how these are always part of our studies, and such naive naturalism/s should be acknowledged as inescapably traveling along with even the most reflective among us. Reading Rousseau is ideal in this respect, because as a method his relation to Enlightenment is ready-made to address our own predicament with and concerns over how science and technology require ethics, responsibility and democracy.

And freedom, let's not forget about how in Rousseau freedom takes center stage, both tragically and in terms of our attempt to expand on the rather narrow origin story that characterizes much of the usual ethics of appreciation – of the multiplicity and newness, of objects, subjects, new technological worlds and their inner workings. With reading Rousseau, comes an uncomfortable sense of urgency, as we recognize that giving meaning to freedom inevitably implies that our own legitimacy as commentators on science in society, on technology and its role, will be at a premium. Our own legitimacy is tied to our understanding of science, technology *and the nature of authority,* as the subtitle of the book puts it, and it implies that we are complicit to the extent that our methods reflect our denial of the tragedy, of the loss of freedom.

Outline

This short book is part of an overarching effort to establish a meaningful sense of interdisciplinarity in-between political theory and the social study of science and technology in its variety. This is a rather large task, given the size of such fields and the differences in their style and character, and this will be no different when focusing on Rousseau.

Rousseau has inspired intimidating libraries of (mis-)interpretations for centuries and there is no shortage of recently published work on Rousseau today. Nevertheless there are only the occasional glances towards science and technology. Usually this is either political theory applied to geopolitics, as a more or less indirect template for various types of international relations theory (see Wendt 1999, Bull 2002, Williams 2005). There are a few works that make extensive use of Rousseau for their techno-optimism, for example celebrating the altruism of human nature in the context of open source software (Benkler 2011) or how with information technology we could realize Rousseau's democratic dream (see Azuma 2014).There is also his influence on literature and pedagogy. Rousseau's work comprises a strong claim on giving shape to the modern novel (Santo 1994, Roulston 1998) and is quite frequently discussed within the context of critical pedagogy studies (Wain 2014, Lindsay 2016, Petrovic and Rolstad 2016, Lilja 2018).

These various types of engagement with Rousseau's work are each relevant when it comes to the commitment to "freedom", as an enlightenment and modernist concept that retains its meaning in a world of science and technology wherein it is at stake. This is why the subtitle refers to "the nature of authority", as the relationship of power and knowledge is changing drastically in its relationship to the sciences and the pervasive role of technologies in our lives. What that implies is that there's no hiding in the past, clinging to a belief that freedom cannot be renounced, that it is inalienable and therefore somehow given to us by history. We experience constantly how it is invariably lost in the smallest of decisions we all take all the time, how freedom in its relation to software often appears as archaic and barely tied to reality of increasingly digitalized lives. Such complacency is a tragedy, and that's why we should return to Rousseau.

The subject area and the overall argument about political and social theory were also addressed in an earlier book, "Recoding life", written together with Sakari Tamminen (Tamminen and Deibel 2019). This short book could be read as a companion piece but also independently as a more condensed application of its methodology. Its objective, once again, is not to establish a causal or deterministic historical relationship, to be comprehensive in a theoretical or empirical sense, nor to prioritize any type of need to produce novel insights into political theory based on close textual reading. Rather, a basic understanding of the depth and range of early modern thinking is needed as a method whereby to establish and approach the object of analysis. Even though (his) words change meaning over time – as did the many conflicting interpretations of his work – a fairly stable categorization of his basic formula is possible. In turn, this book, like the earlier book, combines this commitment to a critical reading of the principle categories of early modern political theory with an interest in a range of contemporary empirical concerns and the social theories woven into analyses.

Accordingly the three parts of the book can be read sequentially but with as a cautionary note that they are somewhat different stylistically. This is the result of directly addressing different types of scholarship and readership. Chapter 1 extends the discussions of the stereotypes that accompany Rousseau and makes the observation that Rousseau's tragic conception of freedom is characteristic of modern social theory in its relation to science and technology. Chapter 2 engages more directly with STS and its relation to political theory while Chapter 3 takes a more polemic tone (it's Rousseau after all), as it seeks to simultaneously develop and apply Rousseau's concepts.

The point of departure (in Chapter 1) is to confront some of the basic misunderstandings of Rousseau and to elaborate on the relevance of Rousseau's tragic conception of freedom. Hereby we sharpen our focus: the need to return to Rousseau as compared to more recent theorizing. Subsequently (in Chapter 2), the discussion of Rousseau will be held in political theoretical terms. At this point the focus shifts to his critique of progress and his conception of knowledge. In many ways his philosophy of science is rather pedestrian today when approached either in terms of 20th-century philosophy of science or STS. What matters most, therefore, are its implications, as can be developed by drawing out the parallel with Hobbes, as the antagonist of Latourian STS and its relation to Enlightenment. Turning to Rousseau's critique of progress, instead, suggests an approach that is at once compatible with and critical of what are today conventional approaches to practical and pragmatic ethics, responsibility and the authority of science and technology in modern democracies. While it is beyond the scope of a short book to deliver full chapters organized around empirical observations (this is what the earlier book did), the goal (of Chapter 3) is methodological in how it seeks to make tangible the suggestion that there could and should be a re-orientation in our critical thinking about freedom. Following a discussing of Rousseauian ideas about the social contract and the general will, these concepts become the focus of a critical re-examination of the relation of freedom and software. While this does include a discussion of "how" to do case studies, these are methodological references and their goal is to sketch out the contours of a more applied perspective.

In this regard a few words of warning are also in place when it comes to free software and open source. These are topics that are discussed seriously, and at some length, particularly in Chapter 1 where they are introduced and again in Chapter 3. However, its point is not an evaluation of the legacy of free software in any practical sense nor its hold over "the future of freedom". What is of consequence is that freedom was raised as "the" problem in the early 80s and in this sense the origins of free software remains remarkable as a response to the extension of intellectual property at the time. Particularly its perspective is important because of how it sought to place freedom and contractualism at the center when intellectual property was extended to source code. This, however, has become more challenging over time, beginning with the rise of Silicon Valley, the influence of intellectual property on the information technology coming out of it and how the value of openness was incorporated into the general technological culture that came to surround it (see May 2000, Boyle 2003, Sell 2007, Berry 2008, Delfanti 2013, see also Deibel 2006).

Finally, it should not be surprising that our end-result will still be mostly tragic. This can be seen as a call for a more "critical" type of STS with a greater recognition for the inescapable role of theory and conceptual work. This is not a rejection of careful and detailed empirical studies but about the methodological implications of Rousseau's tragic naturalism for doing such work. It implies that empirical observation should be done by acknowledging the confines of an ontological and a normative commitment to an outcome, which in strict term is freedom but taken within a wider setting refers to any comparable type of political theory and political philosophy. This is a conclusion that was given in the opening sentence, in that we should not shy away from recognizing the impossibility of our situation (as critique) but simultaneously engage directly with our own predicament in terms of the various and precarious relationships of science and technology to nature and modern society (as STS). The paradoxes and contradictions that Rousseau's writing is often charged with, are part of our reality and in that capacity nothing to shy away from. He didn't and his critical method still holds: our own historical moment requires of us that we move forward, identify a new potential to speak up about freedom, to re-imagine it as a meaningful category and, with immodesty and care, act upon our chains.

1 The Future of Freedom

The freedom-machine

Consider, for an instant, the familiar story about how hardware was king until the late 70s, when software began to reign supreme. We could rehearse the specifics of this origin story at length and in a familiar shape and form. It's not very original, as it would simply involve detailing the conditions that characterized the rise of the software companies of the 80s, like Microsoft and Apple and the struggles of those that preceded them, like IBM (see Stephenson 1999). Let's, instead, simply agree that there used to be a realm where authority had depended on control over a few rare and luxurious machines that were hard to operate. In its successor state, these machines became a bargain and were much easier to use. These new things were more personal, inside your home; however, they mostly ran a multitude of poorly realized designs.

The familiar story revolves around how the new rulers ruled like tyrants, imposing new rules and setting the new norms. This was unlike its predecessor who, in the 70s and before, had still ruled with some restraint, being mostly interested in hardware, software was mostly left free. And just as the sustained attempt began to be made to take away the freedoms of its subjects, and just when it began to look like a hopeless condition, a sliver of hope appeared. Instead of hopelessness, there was the unlikely revolt of a small band of renegade coders. Unknown to the general population, through cleverness and bravery, they challenged the new rules and successfully began something different based on a simple deal. The proposal was almost common sense: you are free as long as your freedom does not result in restricting the freedom of others. To put it in terms more applicable to our own situation, there would be a simple license that was free of all the "user–restrictions". Instead of clicking "I agree",

DOI: 10.4324/9781003193609-2

you would be agreeing to a *"U-R-Free"*, as it were, with as its main condition that you agreed to keep whatever was being shared with you free for others.

As unlikely as it seemed at the time, it was only some years afterwards that the contract turned into a basis of an entirely new way of doing things. The few participants that had joined early on, it turned out, had opted into a new type of social pact that had created something spectacular: a machine that was built by them and made available without any of the restrictions having a hold over how it could be used. Surely this was a machine of freedom, one that with ever new contributor was expanding the domain covered by the contract. Could it be that this was the beginning of the transformation of the kingdom into a realm of associated and yet unrestricted free individuals? One might be tempted to consider such a machine a *"freedom-machine"*.

We know, of course, what is the most obvious candidate to live up to such an imaginative title: it is Linux, or GNU/Linux, as the advocates of the Free Software Foundation (FSF) call it. The latter are referring to a contemporary history that values their groundwork, continued work, as well as to their logo (an image of a gnu) and the relevance today of their principled stance on what making available code should be about: freedom. Key for us, however, is not the rehearsal of an exact and detailed modern history of GNU, the origin of Linux or the many other examples that might be described in terms compatible with "the freedom machine". Rather, such a "freedom-machine" refers to an understanding of freedom, arising from this background, that has become an inextricable part of industry standards today wherever software development matters. To be more precise: freedom-machines run nearly all supercomputers, servers and the basic functionality of smart devices even though its connections are managed by businesses like Google, Facebook, Twitter, Amazon and so on. In fact, each of these mega-corporations, like anyone else with a need for serious computation, runs Linux for themselves.

And it is exactly in this state of affairs that the rhetoric of freedom and its relation to software (that wants to be free) begin to dissolve. Our problem is not that the desire for freedom did not deliver, as its skeptics predicted for decades, or that it is somehow not relevant to the practical concerns of serious business and industry. This much the modern history of the renegade band of coders has persuasively established. Our problem is a different one: the "freedom-machine" did exactly what it was intended for.

Bits, genes, and (freedom-)machines

The freedom-machine is getting under our skins quite literally; skin – as the outer physical boundary of our bodies and embodied individuality at once serves as the boundary site for inscribing new attributes, capacities and dreams into our bodies (Tamminen and Deibel 2019, 34–7). Each of these is powered by the freedom-machine and characterized by a profound capacity to reach out, extend and redistribute our sensing capacities through information technologies transforming into what McLuhan described as an "extended nervous system" (McLuhan 1995).

McLuhan's insight lies out for us the inescapable chains that bind us in a remarkably early modern naturalistic and distinctly Rousseauian vein. It is our nature that is at stake in modern society as:

> Electromagnetic technology requires utter human docility and quiescence of meditation such as befits an organism that now wears its brain outside its skull and its nerves outside its hide.
>
> (ibid, 57)

Such a loss of self, as an independent being within their skin, can quite easily be understood as a self-evident contrast to the Rousseauian ideal of natural men that are inherently free. In its place have come our capacities to understand, communicate and act as increasingly mediated through information technologies (see Tamminen and Deibel 2019). Accordingly, we might foreground such "body politics" whenever we find instances of the historical disciplining of our bodies. Applied to our own predicament, these types of information technologies are extending and intensifying the many interventions that preceded it, like the interventions in the rate of birth, longevity, sanity, illness, reproduction, living conditions and so forth (Foucault 2003).

Of course, there are many such 20th-century thinkers that we could turn to for such an account of the loss of freedom. Each of these could help us establish our point: that there are postmodern lines of thinking that are highly compatible to Rousseau's early modern understanding of how freedom is invariable lost. True enough, it is conventional to address the Rousseauian "dependencies" of modern societies in terms of the sophistication of 20th-century social theories – rather than by returning to early modern political theory. Accordingly, the general point is that there is nothing particularly novel about postmodern theories, at least the novelty was already established in modern and early modern writing. This was demonstrated already by the references above to McLuhan and Foucault. Similarly, when it comes to "freedom-machines" we could turn to many suitable examples, like the many types of abstract machines of Deleuze and Guatarri. But such perspectives (as well as the many other post-modern thinkers that could be discussed along similar lines) take us exactly to the same conclusion but not necessarily any further (see Deleuze 1987; Baudrillard 2000).[1]

Surely it must be true that the Promethean chains that bind us, and that Rousseau invoked, take on a different shape today? Our chains are no longer the same as those that come with a conventional application of political theory to questions of sovereigns, governments, citizens and their boundaries (see Ohana 2017). And yet to invoke a world that runs on freedom-machines represents exactly such a predicament; it illustrates how even a remarkably successful attempt to use software in order to act on our desire to be free ends up as an essential part of a tragic condition that is only post-modern to the extent that a flood of new topics are inescapably matters of concern. At the very least, we should refrain from jumping to the conclusion that our tragedy is best understood as an exclusively post-modern condition. On the one hand, we find ourselves clearly in a post-modern condition in the sense that any commitment to freedom needs to be established on a foundation of increasingly unstable configurations of life, code, facts, data and artifacts that are erasing conventional boundaries between natural and artificial, living and non-living (Vermeulen et al. 2012). On the other hand, our thinking about operating systems, browsers and apps in terms of freedom implies a return to nature that is as much "early modern" in its appearance as it is "post-modern".

The basic position is this: that taking the complexities of post-modern theories at face value has us running in circles. It keeps us poking at how deeply tragic our situation is, revealing to us the fiction of man as "free agent" and how this figure is being comprehensively turned into a technologically mediated and disciplined object of technologies of power. We, the analysts, are caught along in the loop we are describing, a loop wherein different ways of seeing and relating are constantly co-opted. Rather than digging into this point even further and showing these theories in all their nuances and differences, the argument made here is rather that we have in front of us an age-old modern question (rather than a post-modern one). The question in front of us is once again exactly the one that Rousseau was dealing with: how to save freedom from the chains that are proliferating and strengthening? Of course, we will require a little patience to adjust Rousseau's insights and the historical moment they arise out of to how we're dealing with similar concerns in contexts wherein the (im-)materiality of science and technology undercuts our options, closing down the familiar forms of action, organization and thinking.

Fortunately for us Rousseau was well-aware of how science is a part of society, how it exercises authority in its claims on what counts as knowledge and truth, and how power is exercised through authoritative descriptions of nature. Delving into his works will offer us a standpoint that remains largely underappreciated, outside of the limited sphere of highly specialist literature in political theory. Before getting there, however, the next step is to remove some of the more pervasive misconceptions about Rousseau. Most notably Rousseau is often being characterized as unapologetically optimistic in his descriptions of human nature. This includes science and technology with Rousseau as an advocate of altruism and cooperation, effectively turned into a strut on which the citizens' active involvement in the everyday affairs of a free society are to be guaranteed.

Linux as iconic for altruism

The commentary on Rousseau's origin story about human nature is extensive and detailed, and, as his political anthropology has a crucial role to play within his overall perspective, it should be read with interest. However, its relevancy should not overshadow that the details, the empirical dimension, is secondary to the argument; methodologically speaking, it does not matter what humanity's "natural" characteristics are.

Rousseau's descriptions of human nature go in every direction, with sentences for each flavor, which makes sense because man is not presented to us as a singular, essentialistic, figure or trope. There are many sentences that might demonstrate his "passionate affirmation" that man is naturally good (Roosevelt 1990, 25) or that nature provides men "with more goods than they needed" (Hoffman and Fidler 1991, xxxix). However, such affirmations are never about any type of primary instinct or any other type of biological justification of how the individual behaves. There is no bond of sympathy binding individuals, but a passive egoism prevails, one that is "indifferent and unconcerned" to others. Any capacity to be good is about how mankind "lifts itself" to the idea of freedom (Cassirer 1989, 101–4).

Rousseau's "state of nature theory" was formulated primarily as a response to his predecessors in the doctrine of natural philosophy in the 16th and 17th centuries. Writing a century later, it was Rousseau who thereby reinvigorated natural rights theory, which had by that time was thought to have lost much of its dynamism (Wright 2017). It's this response to his predecessors, and not any type of anachronistic readings from the perspective of those he inspired or antagonized that delivered what would become the "basic template for the revolutionary republicanism of the succeeding epoch" (Wright 2017, 89). This context is important because of how much stereotypes of Rousseau have circulated and continue to do so. This includes how Rousseau's ideas have been extended to projects like Linux wherein the source code is freely available, which are oftentimes described in terms of how volunteers and user communities are transforming innovation. The hope in this case is that Linux, and similar type of projects, will set the modern world free from selfishness, egoism, competition and other such vices. Look at how much volunteers working together without profit motive can accomplish? Look at altruism in action, or in a supposedly more Rousseauian vein, look at human beings are "fundamentally capable of empathy, of possessing sentiments that compel us to act morally, cooperatively, or generously" (Benkler 2011, 5).

Marvelous! At last, the world gets remade out of a desire for freedom – free from coercion, as rights-holders, as autonomous beings and as members of political communities. Alas, such a desire is based on our contemporary preoccupation over the match between the selfish individuals of modern economics and, occasionally, even how the biology of the 90s still retained a certain commitment to "selfish genes" (Dawkins 1989; Dupre 2005, see Deibel 2006). It's as an attempt to challenge the centrality of such naturalized selfishness, in modern economics, that it needs to be stated that users innovate, that they share resources, that they establish their own procedures and so on. The end result is a steady stream of more or less idealistic/opportunistic accounts of collaboration, in fisheries, in the sciences, in industry, in software projects and so on (see von Hippel 1988; Ostrom 1990, 2005; Raymond 1999, Bogers et al., 2010).

There are a few instances that such studies might make a reference to political theory, as a historical background to this contrast between selfishness and altruism. While this is usually just an introduction, trivial to the rest of the study, it should still be noted that there are counterparts in specialist literature that might be found in support of such references. For example, the celebratory attitude to Linux, as a type of social innovation, is described by Yochai Benkler (in the "wealth of networks") by how "Tux, the Linux Penguin, is beginning to nibble away at the grim view of humanity that breathed life into Thomas Hobbes's Leviathan" (Benkler 2011, 3). This "grimness" is subsequently contrasted to Rousseau, as the advocate of altruism. While this is not Rousseau's position, such a basic contrast between them is somewhat understandable given that there is a long tradition wherein "Rousseau is particular is seen as profoundly anti-Hobbesian" (Tuck 2017, 37). Accordingly, there are many phrases in Rousseau's work to turn to when seeking to confirm simplistic tropes about altruistic individuals. We only have to go a couple of paragraphs further into "The Social Contract" for Rousseau to mention Hugo Grotius and Hobbes, presenting his own work as the solution to how these theorists of freedom described humanity in terms of its tendency towards selfishness, greed and violence. Therefore, starting with Rousseau as an "anti-Hobbesian" can easily appear as commonsensical, and, by extension, the success of Linux is explainable in terms of a similar type of naturalized contrast between selfishness and altruism.

Rousseau, in this light, could have been more persuasively staged as a supporter of a perspective that seeks to help us understand that there is a massive transformation that is quickly undermining the self-interested agent of modern game theory and rational choice theory. Of course, this was not Rousseau's concern. He did not provide us with any type of overly persuasive version of a natural foundation for how the world works. He is agnostic about the choice between selfishness and altruism as the defining characteristic of human nature. Man might be altruistic but does not have to be, man can be selfish or not. Certainly, modern society is full of egoists and selfishness, but for that to be the case there is no need for statements about human nature as intrinsically cooperative nor for any other type of decisive personifications of naturalistic naivety, innocence and authenticity.

Such naive returns to nature, to natural philosophy and the doctrine of natural law, are exactly the kind of regression into escapist types of romanticism that we are trying to identify, to make use of in our own analysis. Rousseau's descriptions of human nature offer us sentences for each flavor, for each type of "naive naturalism". However, these do not show that he is inconsistent, contradicting himself, because man is supposed to be and become many things. His view of the "dependencies" that tie us down in modern society would include altruism, as mediating and restraining our capacity to be free. This point – that Rousseau's political anthropology is neither this nor that – is important. It indicates already that his method will give us no ground for deterministic or essentialistic accounts of human anthropology or biology and this extends to his critique of progress, as we will see.

There might be some who would expect a more prominent role for Rousseau's turbulent life or his personality traits, to demonstrate that

there are deeply entrenched biases and that these are somehow decisive in our evaluation of his overall argument. These instances, and
they exist, are only useful to the extent that we are certain that our
own personal behavior and our views today will still hold up as consistent, either in a decade or centuries from now. Certainly, he has
these flaws, his views of how men lived outside of his own world, his
views of the role of women in society and, of course, many of his own
personal relations. As informative as these might be, they have no
decisive bearing on his method and, of course, we are free to try and
learn from his mistakes.

Selfishness

Famously, Hobbes referred to life in nature as brutish, nasty and
short, a sentiment which is echoed by Yochai Benkler when he says
that Hobbes laid the groundwork for humans as "fundamentally
and universally selfish" and for the state as having as its purpose
that we do not "in our short-sighted pursuit of self-interest destroy
one another" (Benkler 2011, 4). Were this somewhat accurate, why
not agree? However, it's not accurate when it comes to Hobbes.
The conventional idea about Hobbes would be that man in nature
is fearful and acts aggressively towards others, while a closer
reading would suggest a more nuanced characterization, one that
sees the natural law of self-preservation in terms of men that are
"fundamentally self-protective and only secondarily aggressive"
(Tuck 2017, 38).

  Nonetheless, it might still make perfect sense to reject selfishness,
as a model for modern society, especially when considered in the context of our own second nature and becoming post-human (see Hayles
1999, 3). By all means, let's call out such "possessive individualism" for
what it is (MacPherson 1962). For example we might designate the pro-
ponents of selfishness as "IP-droids" and "econo-dwarves" who have
"retarded progress" – as it was once put by free software advocate
and law professor Eben Moglen in his classic article, called "Anarchy
Triumphant: Free Software and the Death of Copyright" (Moglen
2003). We should have all joined in with Moglen who brilliantly went
on to lampoon the idea that nothing good ever gets created without
proper financial incentives. We should have listened, and today we
should continue to show wherever we can that those that defend such a
theory of invention do not have the history of technology on their side.
Better late than never, let's rid ourselves of the "crummy" metaphor of
the incentive – a society of donkeys following a carrot on a stick that

represent whatever the rules designate as property and the potential object of ownership.

The problem, however, is that the optimism has not aged well. It's 15 years later – after Moglen wrote this delightful text and its hopeful view of the future of coding and its relation to freedom seems like something that belongs to more innocent times, as a distant illusion showing what might have been. What was not at all regrettable were his efforts to lampoon the intellectual poverty of the theory of incentives and its associated greed. This has been incredibly successful, as many other critics noticed that there is a type of "second enclosure movement", raising the question what exactly are the types of commons being fenced off (May 2000; Boyle 2003a, Capra and Mattei 2015). Already at the time, however, it was noted that the theoretical contrast was not so clear-cut; that it might be a stretch to refer to the contemporary situation in terms of 17th century England (Harvey 1996; Hardt and Negri 2000, 300–3). This observation remains relevant today. After all, we can run whatever algorithms we can on a wealth of freedom-machines. With this ability come countless arguments in favor of access and openness to information, data and knowledge and our advocacy of the commons changes character dramatically (see Deibel 2009, 2014).

There is little that is problematic about advocating in favor of the commons when perceived in contrast to the greed of IP-based publishing industries. After all, we're just lengthening the inventory of different kinds of commons, when listing them in the strict economic sense of involving the management of shared resources. Certainly, it's somewhat tragic, a tragedy of the anticommons comes to mind, but the solution is simply the trading and cross-licensing of intellectual properties, the arrangement of access by way of collaborative research projects, joint ventures, consortia, partnerships and so forth. If characterizing these activities as solutions to the anticommons, we should recognize that they represent a considerable strategic and economic value relative to patent strategies. Accordingly, it is an alternative that belongs to an organizational setting of start-up companies, venture capital, stock markets and other kinds of financial speculation, an "overall terrain of cooperation, conflict and value generation" (Sunder Rajan 2006, 58).

Clearly the obsolescence of selfishness is announced somewhat too readily and too categorically. What if the forces that seek to enchain us simply shifted back a gear, affirming the tragedy and the need to go beyond selfishness? There would be one condition, high-tech types of altruism, cooperation and gifting will be at the center of things.

What is "free", as premised on such a lazy contrast, is exemplified by our "freedom-machines", invented together, compiled and maintained by volunteers, users and activists alongside industries that are hugely beneficial because of the ability to generate and capture profits. Accordingly, we might end up trying to "prove" that Linux is based on altruism (with Benkler) or we might prefer to state that self-interest is really the motive of its participants (see Raymond 1999). Such empiricism, however, comes with a claim on what should count as reality and what as evidence, and occasionally it even ends up in lawlike statements (e.g., Raymond's "with enough eyes, all bugs are shallow"). Switching to a more political theoretical register we can easily establish that there's little to gain from simply choosing between one of these two options. If we blame altruism and cooperation, we end up revitalizing the centrality of egoism in economic life.

Regardless, our attention remains with the intricacies of the various observations on, and calculations of, economic life. Whether it is egoism or altruism that we prefer rational choice theory holds – but only as long as we consider them as the only options, diametrically opposed views of human nature that are somehow outside of history. Such contrasts, however, are modern myths and there's no turning to Rousseau to affirm the contrast, that it has a solid basis in political theory. In the closing passages of *Emile* – Rousseau's educational fantasy/political treatise – he famously observed:

> "how little sensible men have read or understood" when Hobbes is "overwhelmed with abuse."
>
> (Rousseau [1762] 1993, 505)

This certainly applies to Benkler's complacent interpretation of Hobbes and Rousseau, which ends us up with a naturalized psychology that is not found in their work. For such a position there are some other places we could have turned to, like "the uncritical enthusiasm with which Diderot praised the state of nature as a state of innocence and peace, of happiness and mutual good will" (Cassirer 1989, 102–3). Alternatively there is Samuel Pufendorf's state of nature theory, characterized by a much greater degree of social relations (Tuck 1999). There might, however, be other problems, considering how these perspectives imply being critical of expansionist business practices in colonial contexts, even if they are motivated to a degree by altruism and social in nature.

To overwhelm "Hobbes with abuse" is to say that he is being invoked as a straw man. For example, when his political anthropology is presented as the foundation for everything that's wrong with modern economics. At best this would apply to the type of modern capitalism that is prevalent in those societies that revolve around singular sovereign powers that impose their authority and their sense of order as much as conceivable. As a normative stance his would be something, at least, but as a liberal position the problem remains. After all most major liberal theorists dismiss such state of nature theories out of hand, as static and defunct juridical devices whereby to specify the terms of social contracts (Rawls 1999; Kymlicka 2002, also Hume 2011). No need to take Hobbes or Rousseau seriously on that front – and not so long ago the specialist literature on Rousseau might similarly have downplayed any substantial similarity between the two and their state of nature theories (see Tuck 2017, 41–4).

There was never a time, throughout the centuries, that the proximity of Rousseau to Hobbes was not a matter of discussion. Certainly, Rousseau's critique of Hobbes did not turn them into opposites, which was well established as soon as his major works were released and "until well into the 19th century, it was customary to stress (and often to deplore) the similarity". This again became the established reading in the middle of the 20th century and this continues today (ibid, Cassirer 1989; Arendt 1997). Its implications are what should concern us. Most obviously both Hobbes and Rousseau see modern society as a condition wherein we give up our natural freedom, although Hobbes sees this as a desirable outcome whereas Rousseau sees it as tragic and inevitable. Nevertheless, this is already an indication of how Rousseau sought to be more "Hobbesian than Hobbes himself", taking over "the most frightening aspects of Hobbes" (see Althusser 1988, 100–1). Instead of characterizing nature in terms of the state of war, it is a description of modern society, as an existential condition with only one solution that is highly unlikely: Rousseau's social contract and no other type will do.

What sets Hobbes and Rousseau apart is how there's no actual exchange or deal between free individuals; the only way to transcend the state of nature is as "a moral totality" (ibid). Like in Hobbes formula, this implies a concept of sovereignty that is absolute, it is a full union of individuals. While with Rousseau the individuals are expected to join freely, they are still contracting with themselves only and not someone else, which is to say that they once again give up their natural freedom and their own interests. Chapter 3 will take up this point further, as a political point about social contracts. Next, however, it is crucial to establish how the proximity between them includes their views of the sciences.

Both Hobbes and Rousseau share a sensibility about power and knowledge. The key, here, is to consider whether there's a substantial difference when it comes to "truth" and its relation to nature. What, in a more recent vernacular, would be Rousseau's take on the making of truth-claims? How can truth be socially constructed, a characteristic of society, and yet it is simultaneously part of a process of de-naturalization? This could be about the nature of fact, of law or both; at this point we should simply seek to understand how he sought to explain such a "chain of inference". Indeed, there might not be scientists with instruments to focus on directly, whether gravitational waves or otherwise, but we can still observe the part that such chains play within his general framework: how these types of chains "continue through social space time until reach social agreement" (see Collins 2004, 13).

To that end, let's make some observations about the basics: Rousseau's conception of freedom, his critique of progress and more recent ideas about science and nature. Rousseau's critique of progress ends up being rather compatible to contemporary STS, with various lines of empirical inquiry to continue as usual. Such an observation is still important, however, because it allows us to subsequently shift our analysis to his various proposals and the method whereby he approaches these concepts, as integral to our own tragedy and our concerns over a possible way out. Our primary concern is, therefore, not law and the state in its territorial sense, and we have to distance his perspective, stylistically, from more familiar political theoretical interpreters. This is mostly stylistic because the argument will still revolve around the same method, the one whereby "he defined clearly and firmly the specific meaning and the true basic significance of his idea of freedom" (Cassirer 1988, 18).

Simply put, it's the individual who freely consents to the law that determines the character of freedom. This "law" is a somewhat more complex topic but the general message has become commonplace through time – that in a just society "each and every citizen shall enjoy a basic equality of status, dignity and material support" (Dent 2005, 44). Our concern, however, is not to evaluate the historical trajectory running from Rousseau to Kant's society of citizens, which administers law universally and as part of a history of mankind. This book's concern with freedom requires a re-orientation away from this more established historical connection. More practically this refers to the type of interpretation that revolves around a territorial understanding of law and administration. The emphasis on power in its relation to knowledge is, of course, part of such considerations, so it is only a certain distance from the more familiar application of Rousseau's thought to countries, or comparisons between them.

In this sense his tragic conception of freedom allows, methodologically, for a quite different relationship to the social contract, as a more general reference to contractualism that seeks to avoid the interest of a person or a group taking over everything. Such a vector implies a different trajectory, one that establishes its critique of selfishness on different grounds and in how it seeks to approach our own imperative, which is to point with regret to the sciences and technologies that keep being introduced with hope – like the freedom-machine – even as they keep failing us, and even though we had expected them to be on our side, to keep us free, to liberate us.

Note

1 There are a few other popular topics that involve notions of the freedom-machine that are not somehow tied to free and opensource software. Aside of Deleuze and Guattari's concepts, bicycles are often described as freedom machines particularly in the context of mobility as part of women's emancipation in the late 19th and early 20th century. Also worth mentioning is the work on the short-lived project Cybersyn. This project was an attempt during Allende's government in Chile to reorganize the national economy around cybernetic principles. Using a single 1970s computer they intended to synthesize government planning with factory worker participation. Also this ended tragically, considering Allende's suicide, the suffering that was the result of the political coup that followed and that ended the project (see Medina 2011).

2 Science and Society

> So long as power alone is on one side, and knowledge and understanding alone on the other, the learned will seldom make great objects their study, princes will still more rarely do great actions and the people will continue to be, as they are, mean, corrupt and miserable.
>
> (A discourse on the Arts and Sciences, 1750;
> Rousseau 2003, 28)

Critique of progress

The quotation shows the proximity of Rousseau to Hobbes. Characterizing the people as being "mean, corrupt and miserable" echoes that iconic sentence wherein life is described as "solitary, poore, brutish, nasty and short" (Hobbes 1985, 186). Neither description is particularly desirable and each refers to a state to avoid being in; obviously, however, Rousseau is describing society and Hobbes is characterizing nature. The difference revolves around the relation of knowledge and power and, for us, "how" this relation operates matters as it mirrors one of the more established concerns in STS.

The famous phrase by Hobbes is preceded by a list of what man lacks in nature. Life is miserable because there's no industry, agriculture, navigation, importation of goods, no architecture, no transportation and no maps. It is not the case, however, that this means that there is no critique of the sciences when it comes to life in society. Hobbes mistrusted (with relentless consistency) any type of society, friendship, trust, association, shared passions, charity, equity and so on. This includes the sciences explicitly. In the very first chapter of the "Leviathan", for example, Hobbes points out that any such social relations can turn into potential sources of strength that undermine the stability of society. While in nature men are primarily fearful and act accordingly, in society there is a constant need to take precautions,

DOI: 10.4324/9781003193609-3

anticipate, subdue and "distrust others by all possible means" (Tuck 2017, 38). Hobbes, therefore, remains skeptical of professional and political elites, who may be corrupted and seek to control others, even challenge the legitimate sovereign.

It might, therefore, be tempting to expect that Rousseau simply "inverts" the Hobbesian formula. This means that Rousseau's man lives in a society where sciences are everywhere, surrounded by others, leading a miserable life without any virtue. This appeal to virtue is, today, seen as an echo of Montesquieu, who famously wrote about the virtues of the ancients, albeit Rousseau transformed this argument into a much more pointed argument (see Kelly 2012, Wright 2017, 67). Nevertheless the similarity with Hobbes remains striking when it comes to how, throughout his work, scientists are continuously characterized by their dependency on others and on everything in society. Constantly the truth about nature is slipping through a scientist's fingers, as it depends on everything and everybody, to the extent that he refers to the learned societies as "schools of falsehood" (Rousseau 1993, 200).

This is already the case when looking more closely at the quotation above. It comes from Rousseau's discourse on the arts and sciences (below: the First Discourse). In it he presents a long list of historical examples of civilizations, from antiquity to his own time. Each of these shows how the sciences and arts do not encourage virtue, either in society or in its people. In other words, progress does not lead to virtuous people and therefore civilizations fall. The examples speak for themselves: "take Egypt, the first school of mankind", "take Greece, once peopled by heroes", "Rome, founded by shepherds", "What shall I say about that Metropolis of the Eastern Empire, the Chinese should be wise, free and invincible" etc. (Rousseau 2003, 8, 9).

The First Discourse ends by expressing the hope that the "princes" might someday be able to show what "virtue, science and authority can do". This would only be possible, it is concluded, if they listened to the advice of "learned men of first rank" (he mentions Bacon, Descartes and Newton as examples). These are the ones that make great objects their subject of study while their wisdom is matched by their ability to be critical of fashionable norms and commonly held beliefs. Without such advisers and princes that listen to their advice (as the quote at the top of this chapter reminded us), the people will remain without virtue: "mean, corrupt and miserable" (ibid, 28). However, the First Discourse does offer us a few remarks that are not critique. After having established his critique, he states: "Let us consider the arts and sciences in themselves" (Rousseau 2003, 14, see also Howard Campbell and Scott 2005). While this is only stated in closing, and in contrast to the lengthy and relentless critique that precedes it, it is already an early indicator that with Rousseau, knowledge is not simply power.

The type of critique of progress that Rousseau delivers does not revolve around any type of monolithic or singular rejection. Instead, his critique of the sciences and the arts concerns the "high esteem attached and time devoted to them". This echoes many contemporary debates, which in the vocabulary of ethics and responsibility are concerned with a lack of "honesty, integrity and service of the common good (Dent 2005, 51). Therefore Rousseau's suspicion of the role of scientific authorities, their claim on truth, is not simply critique, some type of perspective wherein knowledge is nothing else than power, as if somehow they could be a singular category without any distinction between them. This is often projected onto Hobbes (Latour 1993, 26, 2017). While his theory of the state deservedly has this historical reputation, it is somewhat of an exaggeration when it comes to knowledge.

It's hard to pin on Hobbes a perspective wherein "cognition is wishful thinking all the way down" and truth is "no more than just a happy accident" (Collin 2011, 59). He was an accomplished geometrist, influenced by Descartes and a product of a humanist literary culture, even if he turned away from it. Accordingly, Hobbes is mistrustful of scientific reason's powers to persuade people, carefully (re-)considering the humanist assumption that the findings of reason need to be supplemented with the force of eloquence (Skinner 1996, 217). Obviously Rousseau had no such reservations, his rhetorical gifts are there for all to see, while Hobbes makes for dense reading. Otherwise, Rousseau's mistrust of the sciences echoes that of Hobbes as a line of critique as well as our own concerns over science in society. The objectivity of facts, their methods and social organization of science are increasingly under pressure. The sciences, in such accounts, are essentially furthering their own interests along with expert systems alongside the "world of fashion, film, popular music, television and writing". Each of which can confirm how relevant "Rousseau's assessment is of what is going on" (Dent 2005, 55).

Maybe, however, Hobbes' mistrust and Rousseau's skepticism of the sciences are simply unremarkable, not at all shocking to today's audiences and through how well-established it has become to challenge the special status of the sciences, its authority as based on facts and a privileged relation to nature. After all, the status of observation and experiments, as tied to a rational order of things, invariably comes with a type of naturalism. Therefore science is a "vehicle of naturalization" and placing it in a historical context implies an approach that does not leave it a special space, somehow separate in its relation to knowledge from the "hustle and bustle of society" and with a "place of special authority for science vis-a-vis ordinary societal concerns" (Collin 2011, 82).

Accordingly critique, of the special status of scientific knowledge, is nothing new. Hobbes and Rousseau's discussion of the sciences shows that this type of critique was integral to "Enlightenment" and to "the Scientific Revolution". Considering the status of both, within political theory throughout the centuries, the observation might even be that critique predates the later affirmations of progress or the positive role of the sciences in society. This should change our understanding of the "post-truth" world that we might be living in. If "truth" somehow is seen as a characteristic of the Enlightenment, so is its critique. Hobbes's critique of the sciences and Rousseau's critique of progress show the extent to which it was part of Enlightenment to observe that "truth" cannot be turned into the foundation of social order. If our concern is that modern times have left behind the Enlightenment commitment to science and truth, the realization must be that this happened already at the time of its arrival centuries ago. If anything at all, our concern over post-truth should instead be about how truth and critique go together and only now we are leaving their connection behind.

This initial observation becomes important again in the next chapter but for now it is sufficient to state that with the First Discourse we can already see that Rousseau sees the relationship between power and knowledge as interwoven and interdependent. His later work will make it much more layered, which is where we can establish its relevance in parallel to Hobbes, in line with the latter's rediscovery in STS and indicative of our own predicament with the state of our thinking about science and technology in social and cultural terms.

Hobbes and social constructivism

Maybe the suggestion that Hobbes has a commanding presence in STS is an overstatement. Certainly there are few lists of "classics" that do not include him, suggesting that he has some type of status within the curriculum of STS. Yet most of the interest goes to Robert Boyle, his counterpart in Shapin and Schaffer's Leviathan and the Airpump, from 1985. Boyle is generally remembered for the law of gases but Shapin and Schaffer's study shows Hobbes as the challenger of Robert Boyle and how this conflict was crucial for the nascent laboratory and the experimental culture around it. Again placing the focus on the part played by Hobbes, rather than Boyle, will therefore go some way in foregrounding the detachment of influential strands of STS from any meaningful relationship to political theory.

This is not a problem in Shapin and Schaffer's book on Hobbes and Boyle, which prominently has an adjusted version of the Leviathan on the cover. Instead of the staff of religion, there's an airpump in the hands of the king. Accordingly, this study of "the experimental life", as the book's subtitle puts it, studies Hobbes alongside Boyle as he disputes "how authentic scientific knowledge should be secured" (p. 3, 4). Hobbes' presence, therefore, is not optional, an add-on to a story that is really about Boyle. The authors argue this explicitly: Hobbes' objections were believable and there was "nothing inherent in them that prevented a different evaluation" (p. 13, 14). Therefore, there's more to Hobbes' critique than that he mistrust the politics of these types of experiments, the practices and the conventions that came with them. Tellingly, he also rejected the experimental relation to nature on its own terms. He challenged the premise that human observers would act as neutral witnesses to the establishment of facts within the confines of experimental spaces – pointing out factual errors, inaccuracies in procedures, obvious human biases and exaggerated claims attached to the results (Hunter 1994).

The overarching point is therefore that there was nothing self-evident or inevitable about the outcome: a natural philosophical consensus in favor of the experimental program. As Shapin and Shaffer argue, they had demonstrated that there were two competing solutions to the problem of order in society. While Hobbes lost, this was not given, and the last line of the book controversially even argues that the controversy showed that "knowledge as much as the state, is the product of human actions. Hobbes was right" (p. 344). It is this line that Bruno Latour picked up on, in *"we've never been modern"*, arguing over what this means; can Hobbes be right when it is he who rejected a relationship of knowledge to power wherein there was little to no space for experimentation? (Latour 1993, 1999, 2017). What matters is this Hobbesian relationship between knowledge and power, one that is at once identified as the root of "Realpolitik" and as the inventor of the "basic vocabulary" of sociology, its modern meaning (1993, 26). This includes Shapin and Shaffer, who are told that, in their last chapters, they ended with "an asymmetric way of explaining knowledge through power". The use of words like power, interest and politics, "as the only valid source of explanation", leads us to miss out on the process whereby we ended up with the "modern" constitution (ibid, 29). Latour again includes "nature" in his analysis, rather than relying on social categories to overcome the dichotomy. It is still nature, but without any "higher epistemic levels" nor any "deep epistemological foundations", like knowledge and a state of nature theory, respectively (Collin 2011, 134–135).

The stakes are high, as can be understood from how John Law refers to this case. In a text called "STS as method" (2015), he begins by explaining that STS rolls theory, method and empirical practice together. After stating that STS works through its case-studies (line not added), we go from the origins of STS to the moment that nature was separated from the social (this is case study 2, line added). Hobbes is never mentioned, as we are told that what matters about the case study is: 1) the air pump experiment as a material technology, 2) the writing up of "facts" as a literary technology and 3) the selective group that was legitimized to act as witnesses, as a social technology (see also 2008).

On the one hand, this captures a sensibility in STS about how theory is to be done and the status of concepts (like power, politics, interest, etc.). Accordingly, "it's the empirical case study that are important" and STS fits "uneasily with stronger, more foundational, and more critical sociologies". STS, in his version, is character-ized by "its refusal of grand narratives and the macro" (Law 2008, 630–636). This is, on the other hand, not an uncontested description of what STS is, or should be. Söderberg describes Law's definition as "post-structuralist" STS and points to Law's description of this posi-tion as "an amalgam of French epistemology and English empiricism" (ibid, see Söderberg 2017, 201). In that capacity, it is compatible with how Collin refers to Latour's philosophy as "radical ontological mon-ism" (Collin 2011, 128). At stake is not a choice between empiricism and overarching explanatory frameworks and elevated vantage points (to critique). Rather, the issue is its implication: that in STS there's only this position, doing empirical case studies, from where it is pos-sible to meaningfully question modern dualisms – of technique and politics, nature and society, human and machine and so forth.

Consider how Latour extends the adversity of dualisms to the sub-ject and the object, the actual and potential, the real versus the ideal and so on (ibid). Is this not suggesting an overarching framework? Such a framework, after all, is one that suggests that theory should be restricted to case studies and it is dualistic given that it implies a stark choice between philosophies and sociologies that are invested in such dualisms in the wrong way and those that do so in the right way (see Söderberg 2017). The same applies to the status of Boyle and the air pump as an exemplary case study. Its function as a case is to appear solid enough to suggest a relation to the scientific revolution but without having to theorize it, commit to categories like power, interest and politics (like Shapin and Shaffer did), let alone class, capitalism or the state (ibid, 185). Such a position is typically organized around an "unflinching stance on multiplicity". Söderberg maintains that it is expected that "the ills of the world, or at least all the ills that a social scientist can do something about (without making things worse), stem from erroneous beliefs in totality/necessity/master narratives" (ibid, 200).

Taking a closer look at Latour's comments on the controversy between Hobbes and Boyle shows this problem quite clearly. The way that Latour presents this case, on whether Hobbes was right, begs the question of exactly how socially constructivist our relation to reality needs to be. Latour sees social reality as generated in much the same way as the study of science deals with the creation of scientific facts and the reality with which science deals. In this sense, "entities", irrespective of whether they are scientific, take their conceptual shape only after there is a basis for them to exist. This applies to interests, politics and so on, alongside microbes, clams and air pumps. Invoking one to explain the other would be too tautological, explaining a subject to affirm assumptions about that subject, theorizing to affirm a theory (see Collin 2011, 116).[1]

As mentioned, Latour re-introduced "nature" back into the study of science, and this time around, "nature" is a very different category than the one posited by scientists, like Boyle, or those of today. Regardless of what this different "nature" ends up being, the turn to political theory is a specific one. It is an interpretation of Hobbes's state of nature theory, which allows him to diminish the reliance on social and sociological categories, like power, interest and so on. He reworks the category of nature, disentangling it from the relation to the natural world that Hobbes posits. What if, however, this contrast is not decisive at all? This is what switching back to Rousseau's category of nature can show; it dissolves the suggestion that there is a necessary contrast between Hobbes' nature and Latour's "new" nature. By extension, it can be shown to undermine any notion that case studies are not political theory like those of Hobbes and Rousseau; it would undermine the idea that political theory is invariably caught up in grand narratives and that, by extension, it is not STS.

There is, in other words, quite a bit of Latourian purification going on in STS, which Rousseau might not only have ascribed to the sciences but also to established perspectives on science in society. And this type of purity tends to become a demand, one that new members who would like to be admitted to a philosophical club need to face (see Rousseau 2003, 25). Less polemically, however, we can simply state that it remains fashionable in much of STS to be skeptical of any assumptions about a-priori commitments. While this might be a sensible instinct, it becomes an untenable position when delivered as a viewpoint on Enlightenment and Modernity. After all, the case study was never just about Boyle and his experiments, or even just about the status of experimentation in modern societies. Hobbes, at the very least, is an integral part of the case study and, by extension his state of nature theory and its legacy in political theory are part of how STS let "nature" back in.

Simply put, Latour's intentions aside, grand narrative never went away. It travels along with STS as its relationship to its own (somewhat deconstructed) foundations and re-affirms its presence whenever seeking to theorize and conceptualize STS in terms of its own politics and its own relation to nature. Again, the point is not that there's no place for deconstruction as a method, for a deconstructive attitude to origin stories and foundationalism; simply that it cannot imply a complete a-priori commitment to fully epistemological relativistic accounts of the relationship between power and knowledge. This matters here as it would imply that the remainder of this chapter and this book can only qualify if it continues by keeping to strictly empirical foundations for any historical discussion of Rousseau analogous to the controversy between Boyle and Hobbes.

Such an imperative would be hard to live up to, but fortunately it would also imply a double standard, given the prominent place given to conceptual discussions and the theoretical character of the Boyle-Hobbes controversy (see Latour 1993, 14–26). The same applies to the controversy as illustrative of the gendered character of the scientific revolution (Haraway 1997, 23–27). Apparently there are instances that the proximity to the case can be skipped and so this book can continue in the same vein.

Rousseau, social constructivism and its extension to the social contract

There are always further studies of the Boyle-Hobbes controversy and this includes studies that show that the controversy really did not play out like that. For example, we could present Hobbes as a continuation of an established humanist literary culture and as someone who did have his supporters who were an integral part of the new experimental life (Skinner 1996, cf. Shapin and Shaffer 1985, 131). What happens with calls for more empirical (or anthropological) engagement is that there is always more complexity, such as histories that show exactly this about Boyle's historical situation (Hunter 1994). Alternatively Boyle claims only a momentary place among his contemporaries and within "long" histories of how observation came to characterize the experiment (see Daston and Lunbeck 2011, also Daston and Park 2001). In turn, this type of problem – infinite regress into ever empiricism and its historical complexities – could also be applied to Rousseau.

The political theory of figures like Hobbes and Rousseau typically operates like a specialist field where referencing textual and intellectual sources is what establishes a claim, the interpretation of its "reality". This is, of course, quite similar to STS in its emphasis on empirical cases and methods as means to establish proximity to the natural sciences (e.g. Latour 1999). Such an empirical approach could be done, for example, by making a study of Rousseau's position on the sciences as the lived reality of a practitioner. We might learn, for example, that Rousseau practiced music, geometry, chemistry and botany. Throughout his life he was an active participant, at times immersed in scientific networks and generally an "an integral part of the amateur culture of sciences of mid-eighteenth century Europe" (see Bernardi and Bensaud-Vincent 2012, 60). While it is well-known that he wrote on a wide variety of topics during his time working on the *Encyclopédie*, less appreciated are his attempts to write a textbook on chemistry and a dictionary of botanical language. He also collected hundreds of specimens of plants, carefully captioned, and had a herbarium (ibid, 66–67).

In short, there is always more empirical complexity to engage with, activities to describe in detail, their influence over his work and so forth. What this goes to show, again, is that there's more to his engagement on the sciences than simply the critique of the First Discourse. By extension Rousseau's counter-Enlightenment, his skepticism of the role of the sciences is not thereby a full rejection of knowledge in its relation to power.[2] Specifically a perspective focused on Rousseau's relationship to the sciences would, ultimately, revolve around how his conception of nature is integral to his discussions of freedom, popular sovereignty and the social contract.

Indeed, it could be maintained that this is exactly what Latour did, when he interpreted the controversy of Hobbes and Boyle, as the core of the argument of "we've never been modern" (Latour 1993). What is "modern" (not "non-modern" like STS) is how Hobbes's state demands the "incontrovertible assent from all it subjects", only a strong state will do if people are to stop cutting each other's throats. What is "non-modern" about it is that this state is "impotent" without science and technology. For this to be possible, Boyle's science required precise delimitation of the scientific sphere, from such a state, so that its inventions could circulate and populate the land. What, however, if also Hobbes is more complex than that? What if that complexity includes his legacy in political theory? What if that legacy brings us back to the principal categories of today's sociology?

The overarching point, however, is that Hobbes is more than an incidental side character in a story about Boyle and the delineation of science from society during the Scientific Revolution. In effect, Latourian foundationalism provides an understanding of the origin of the historical boundaries between science and society as constantly displaced and constantly in need of deconstruction. It affirms that they matter, and that they are to be rejected as a grand narrative. Fortunately, however, there's no need to simply ignore Latour's observations, as not-a-case study, set it aside as a grand narrative, a modernist origin story, the objectification of the separation of science and society, which was really more complex than that. That's not needed when we turn to Rousseau, even if such non-existent boundaries are clearly well-established within STS and its courses, as a problem to get rid of, the nature of things that it seeks to transcend.

Consequently the remainder of this chapter will not be a case study, it is not empirical in its relation to Rousseau and does not seek to rescue reality from how the workings of science have been "shrouded in philosophical mystery" (Collin 2011, 83). Or rather, it is not empirical as a constant imperative that is, equally mysteriously, still not a regression to an "old-fashioned empiricism-cum-scientism as metaphysical chic" (see Söderberg 2017, 201). There always are dualisms held high, being brought low or hybridized. This includes how, quite recently, Latour suggested that the Leviathan, Hobbes's theory of the state, is being called into question by a hybrid of geology and anthropology. He even suggests that this particular hybrid should be seen as a social contract; maybe it is this alliance that, at last, changes things. As he asks without answering: "could it be through a new contract?" (Latour 2017, 148). There it is, the social contract, even with its natural foundations, albeit in its sophisticated and paradoxical guise. Earlier there was already Latour's non-modern constitution with actor-network theory as its methodology, which he acknowledges as inspired by a natural contract (of Serres 1995). More recently the notion of (bio-)constitutionalism (Jasanoff 2011) has attracted some following in the wake of Jasanoff's popular concept of "co-production", which refers to the same idea about modernity in constructivist terms, as a continuous process whereby societies form their epistemic and normative understandings of the world at the same time (ibid). In this new incarnation, however, there are little or no historical references to naturalistic foundation, and instead they have been replaced with "bio", which refers to cases discussing cloning, embryos, cloning, forensic DNA and so on.

Clearly new versions of these social contracts would benefit in their articulation from retaining some explicit relation to their historical roots. Their basis in early modern political theories as well their interpretation is a necessary step to more mature representations and more nuanced readings of the intellectual traditions STS is channeling. Indeed, time has been lost, given that these discussions did leave a strong mark on STS in the 80s and 90s. One might suspect the interest in political theory stayed there as a result of the turn towards what Söderberg describes as a post-structuralist position, which distrusts grand narratives and the ability to establish meaningful theories and concepts. Our concern, however, is somewhat more specific to demonstrate that a return to political theory, after some time of serious engagement, might lead to more astute conceptual observations and, even, richer conceptual and theoretically informed empirical case studies.

Rousseau is the most obvious candidate because of the proximity of Rousseau to Hobbes. Proximity does not mean that there are no differences or that they do not matter; quite the opposite, it is exactly because of those differences that we should be aware of how "Hobbes's fundamental ideas strikingly resemble the normal description of Rousseau's" (Tuck 2017, 41–44, see also 1999, Cassirer 1989, 101, Althusser 1988, 100–101, see Arendt 1997, see Wendt 1999). This is important as it shows that Rousseau could quite easily take Hobbes' place at the center of STS's relationship to the scientific revolution, Enlightenment, counter-Enlightenment and, of course, whether or not we have ever been modern, or post-modern (see Latour 1993).

In parallel to Rousseau's relation to the sciences, already Hobbes' is obviously more complex than in Latour's version. This can be shown quite simply by observing that the subjects were not simply forced into assent; they were expected to internalize the moral authority of the state. Tuck calls this his "utopian streak" (Tuck 2017, 48), which is a charge usually identified with Rousseau's influence over the idea that states create the citizens that they need. At the very least the suggestion would appear to be that we might always have been, somewhat, modern. Maybe it is even a modernity that always already had non-modern characteristics mixed in. Accordingly Hobbes might have been right about power and knowledge (Shapin and Schaffer 1985, 344), or not (Latour 1993, 26); this is only relevant if we affirm Hobbes as the one that collapses knowledge and power into one, along with a perspective identified with all of realpolitik and generally undemocratic tendencies in modern society. If he did not, and Rousseau's version of Hobbes replaces the latter's mistrust of experimentation, our view of what it means to be "non modern" needs a different understanding of the role of scientific knowledge and technology "in society".

Consider how Hobbes is presented to us as a hybrid creature, both scientist and politician. His views of science could have won, Shapin and Schaffer tell us. But what about Rousseau's scathing critique of progress? It overlaps with Hobbes' views of science and such critiques did turn out to have their hold over modernity. The question, however, is whether this critique is simply another quintessentially top-down (asymetrical) perspective, reducing the sciences to questions of social order (see Latour 1993, 26–29). Like Hobbes, Rousseau mistrusted the authority of the sciences and by extension the claim today on universalism, inclusivity, accountability, responsibility and so on. Such procedures, however, are elements of a political theoretical argument about popular sovereignty. They were always part of the "original STS concern" with public involvement (Collin 2011, 106), extended today with questions about democratic agency and legitimacy and their hold over scientific and technological decision-making and design. These are among the key topics that establish STS within faculties today, teaching social and natural science students based on the idea that it has the potential to have a lasting and positive impact.

Today there might be more hesitation in regard of the advocacy of popular involvement in science, in contrast to concerns over the status of experts and the authority of science in today's society. Nonetheless, this implies a nuance that is still tied to political theory, particularly when considering Rousseau is not the advocate of direct democracy he is usually made out to be. Consequently there is no reason to dismiss Rousseau, as Latour did with Hobbes, despite their proximity. Just because Rousseau was harsh about the sciences, there is no collapse of knowledge into power and it therefore remains relevant "how" Rousseau's critique of science and technology informs his work on popular sovereignty, which is what sets him apart from Hobbes as the advocate of dictatorship. To foreground such differences, it is necessary to understand the similarities. Soon (in Chapter 3) the differences will be the topic, focusing on what Rousseau is much better known for: his vision of the social contracts and its extension to these types of concerns in STS.

Next, however, it will be shown that his critique of progress is perfectly compatible with most contemporary sensibilities about foundations and essentialistic arguments in the sciences. Such problems are, at its core, already incorporated in Rousseau's method, part of how he discusses his state of nature theory as a process of leaving nature. Once we recognize this, we can consider how to apply this method and look at its implications for STS in terms of interdisciplinary relations to early modern political theory (e.g. the social contract and freedom) and its intellectual interpretation.

Knowing nature

Rousseau's naturalism rejects that the state of nature is an observable realm that existed or exists today, either on the outside of modern society or in direct relation to the sciences. Importantly, given STS's anti-essentialism, this means that the state of nature is neither historical nor biological; he did not present it as something to be established based on evidence and this is quite explicit in the texts:

> The man who speaks of the state of nature speaks of a state which no longer exists which may never have existed and which probably never will exist. It is a state of which we must, nevertheless have an adequate idea in order to judge correctly our present condition.
>
> (Rousseau 2003, 44)

Consequently it's Rousseau who turns the state of nature into a thought experiment. This is more commonly identified as a critique of state of nature theories, by those who got rid of them, like Hume and his empiricism and Kant with his appeal to critical reason. Indeed, state of nature theories had been on the decline for quite a while and Rousseau reinvigorated a tradition that seemingly was disappearing (see Wright 2017). Rousseau's perspective, therefore, is at once a denial of the realism of the state of nature and an affirmation, considering how much work he puts into writing his own version of the type of origin stories that characterize the doctrine of natural philosophy before him.

Accordingly Rousseau continues to consider in detail the conditions that lead man to leave the state of nature and give up natural freedom in exchange for the benefits of a social contract. This type of origin story includes Grotius, Hobbes, Pufendorf and Locke (see Tuck 1999). The suggestion that there is a "real" state of nature, one that exists "out there" had long lost its traction and had come to be dismissed as obscure abstractions without roots in history, reality or circumstance. Once the rule of law and political order got its own history (notably with Montesquieu), state of nature theories quickly came to be seen as superfluous, optional and as opposed to serious commitments to freedom in society. This continues today, with liberal theorists who deny the need for state of nature theories, and sometimes Rousseau's histories are placed at the beginning of historical sociology. In this capacity, his investigations of society are lauded as pioneering and distinct from his predecessors. Nonetheless, they are pioneering only in the sense that they are not empirical enough and have been decisively surpassed by the accomplishments of the next centuries, such as the historical materialism, dialectics and related accounts of history following Hegel and Marx (see Simpson 2006, Neuhauser 2014).

Crucially, however, it is not of decisive importance for the unity of Rousseau's through whether or not his empirical statements are deemed correct. Otherwise we might as well jump to the conclusion that also Rousseau's many naturalistic statements are similarly flawed, biased and lacking empirical accuracy in his descriptions of history, natural history, psychology etc. However, it was not accuracy that Rousseau aimed for when making observations on nature in terms of the pre-history of society, selfishness and altruism, as an idealized space on the outside of modern societies or otherwise. Each of these subjects is discussed within the confines of an intellectual framework that establishes its approach without having to rely on any single set of observations on life and nature as empirically observable phenomena, on capturing a natural world that is essentially knowable through direct observation.

This changes the status of Rousseau's empirical realism, for example when making claims based on the ability to establish proximity to natural scientists or, sometimes, by thereby seeking to persuade them about what reality is (see Latour 1999). We should, instead, move past the many empirical contradictions in Rousseau's writing. He certainly believed in his own consistency and coherency and so did some of his interpreters (see Cassirer 1989). The point, however, is not to establish conceptually whether his thought is entirely consistent but to appreciate how *methodologically* Rousseau was able to investigate how the social world is saturated with naive naturalisms of various kinds (see Tamminen and Deibel 2019, ch1 and 2). As Pagden explains it, the state of nature was a "thought experiment" but one that operated as "a condition that we carry with us wherever we go"(Pagden 2013, 166).

Rousseau's critique of progress, in his earlier works, might arguably be a normal part of modern life. What keeps it relevant, however, is "how" in his later work he sought to extend this critique to the unsolvable "riddle of nature", as it had been discussed by his predecessors. While there are many sides to Rousseau's observations, and to those of his interpreters, his method is still our way into the intricate and layered discussions that took place within the confines of the doctrine of natural philosophy (Virilio and Lotringer 1983, 36, Deibel 2009). Like Rousseau, we are seeking a method that does not only deliver critique, it should also be seen in relation to its ontological claim on reality. In his case it is the concept of "freedom" that is approached as ontological and if we "concern ourselves with its fundamental assumptions", his work "appears rather as a thoroughly contemporary and living means of approaching problems" (Cassirer 1989, 37).

Such an approach incorporates historical commentary and disqualifications, like the ones that Rousseau addresses as well as the various commentaries across the centuries. Criticism was already under way when he was alive and state of nature theories were again depreciated with ever more intensity from Hume and Burke onward. Throughout the centuries Rousseau's ideas were pressed into all manner of extravagant ideological service. This is, however, not an obstacle. Commentaries that point to logical inconsistency, paradoxical statements and faulty empiricism should not be seen as a disqualification but can be considered as useful input for his method. The same applies to observations that point out that parts of his descriptions of the state of nature are objectionable, and not applicable to today's circumstances. These are no different than how Rousseau's polemical critiques of the naive naturalism of his predecessors were gradually incorporated into his state of nature theory, as a basic formula that had to be "conquered step by step" (ibid, 40). Similarly the relationship between knowledge and power can be approached methodologically. Also this focus implies that there are many precarious relations to nature to be identified, each of which is to be purposefully idealized into a "non-lost" paradise (see Ohana 2017, 387).

Once we accept that Rousseau engages with naturalism in a layered and nuanced manner, the comparison with the familiar repertoire of established views on science as an ideal and as a practice becomes rather mundane. Rousseau is seeking to idealize science and its relation to nature but, on the other hand, he would be critical of any attempt to isolate ideal-typical appeals to the scientific method from its existence in society. Therefore, his view of knowledge comes with a strong emphasis on the social status of science.

Accordingly there appears to be a stark contrast with philosophers of science who are seen as committed to formalized rules, methods and models, like Popper and Merton. This should, however, not be exaggerated; appeals to Popper's norm of falsification or Merton's ethos of science would not necessarily contradict Rousseau's position, as it should be clear that both were meant as useful ideal-types only.

Merton's ethos of science was not meant as a realistic description of the social system of science, and Popper's attempt to demarcate science from non-science was intended as a means to safeguard an open society from authoritarianism (Popper 2002, Merton 2017).[3] Merton and Popper acknowledged that the ideal and the reality of the scientific method and the establishment of facts are not mere extensions of one another. Therefore, even if science is somehow special, there is no contraction with Rousseau and this, obviously, is even less the case when we turn to Kuhn and Feyerabend. In this regard, Kuhn's "normal science" might be taken to describe such idealization of rules and methods in science but this is temporary within the context of the Kuhn cycle and its paradigm shifts. Inevitably, whatever is the norm, and its procedures, goes into crisis. As part of the Kuhn cycle, the paradigm will not last, as periodically there will be a comprehensive paradigm shift. Feyerabend even describes modern science as fully blinded by its own mythos of truth and progress. Breakthroughs are not supported by attempts to fix norms and rules across different fields that are widely different in their relationship to their objects of study. Quite the opposite, general demarcations of science and non-science are unrealistic and detrimental (Feyerabend 2010, Kuhn 2012).

There are differences that matter, of course, in how these perspectives end up affirming that the sciences exist in a precarious relation to nature. Clearly the perspectives of Kuhn and Feyerabend are more readily compatible with how Rousseau characterizes modern scientists as tragic figures whose commitment to the reality of things invariably end up caught in a web of procedures (e.g. Rousseau 1993, 200). The point, however, is that none of them acknowledges there is singular or even a straightforward relation between the practice of science and nature, with the latter as a category that can be comprehensively determined, theoretically or empirically. Ultimately the disagreement over the status of ideal-typical descriptions of the scientific method might have its advocates and its detractors but each of these in various ways affirms that the many relations of the sciences to nature are precarious.

And this was, on closer reading, also what Rousseau's approach shows about our basic relation to the state of nature, raising the

question of how to understand its relationship to the principle categories of society, the citizen and the social contract. Or, to switch ver-
naculars, what if we follow "the people" that are miserable because of
how knowledge and power are kept apart (as Rousseau would have it)
back to the chains that we started with? These "Promethean chains"
are everywhere and Rousseau even describes himself as meeting
Prometheus, holding the fire and warning him of "the vulgars", and
how the science and the arts would be seduced (see Ohana 2017, 383).[4]
It is, therefore, not Rousseau who suggests that "we have nothing to
lose but our chains", putting us in the position of Prometheus bound
to the rock (Marx and Engels 1969). Nor is this a version wherein we
end up being concerned about (collapsing this mythical origin story
into) the many "chains of inference" that populate our lands (see
Collins 2004, 13).

The latter come in many forms; there are, of course, gravitational
waves (ibid) as well as "entities such as the fetus, chip, gene, bomb,
brain, race, ecosystem, seed and database", each of which is "partly
like Robert Boyle's airpump" (Haraway 1997, 270). Rousseau's
Promethean chains, on the other hand, would integrate these inferences to its "re-appropriation of presence"; a method wherein the
rejection of progress – as part of the corruptions of society and the
degeneracy of culture – is replaced with various types of affirmations
(Derrida 2008, 50–52). Rousseau's literary works, his return to childhood, his confessions of various kind, each comes "in-the-place-of";
they act as a "supplement". While this was once considered one of
the blind spots of his work, it is highly reflective and, in that regard,
should be familiar to readers of STS that it can be "irritating" and
"maddening" to have to track the multiplication of "supplementary
mediations" that end up operating as a "chain of supplements" (ibid).

Rousseau as an immodest witness on education

The Hobbes and Boyle controversy, as emblematic for STS and its historical relation to the scientific revolution, is also the central theory
and method in Dona Haraway's remarkable book: Modest_Witness@
Second_Millennium.FemaleMan©_Meets_Oncomouse™.

While the book's not at all forgotten, its sweeping scope is still
underappreciated. The text presented many isolated technical areas
(like code and genes) as inseparable from each other as well as from
issues of gender, sex, race and indigeneity. Hereby the book was
able to offer an astonishing foresight (in the late nineties), which has
become less easy to notice. Today, after all, there are many more such

studies, offshoots that specialize in the empirical circumstances of the relationship of the life sciences to genetics as well as their position in culture and their part in how identities take their shape as well as their politics.

Therefore, it's admirable how Haraway's insight required not only engagement with the sciences but also artistic creativity, literary aesthetics and a love of transgression (of boundaries). We should want to keep this approach at the center of STS, in combination with a deepening of our interdisciplinarity, so that we do not shy away from including a wide diversity of theoretical fields into our "theory-building projects" and in "our accountability to freedom-projects" (Haraway 1997, 191, 192). However, when this refers to political theory, as a specialist field, that means we might have to incorporate literature that does not match our own preferred and highly specialized aesthetic and literary standards. Such interdisciplinary work is hard, and it is not helped that there is no sense of urgency attached to it. After all, she argues, it does not matter whether "freedom, justice and knowledge are branded as modernist or not; that is not our issue"; the yearning for such categories is "not about putative Enlightenment foundations" (ibid, 267).

Is this the case? Does it not matter "how" we go about to "pragmatically" undo "the founding border trace of modern science – that between the technical and the political" (ibid)? As mentioned, when you make a critique of historical dichotomies a central part of a critical analysis, they will travel with you to the present, when you are being pragmatic about how to respond to one of the many crises that characterize the present day. Inevitably we regress back into the dichotomy we are critical of, our preferred stances mirroring our starting point. Fortunately such a basic push towards this type of accidental synthesis – wherein we end up being both master and slave – is not necessarily a disqualifying type of remark. Quite the opposite, it should inform us that we need more than just case studies and pragmatic methods for theorizing and conceptualizing. Indeed, in this sense, we should return to Rousseau's method, including the criticisms on his complicity towards euro-centric and colonial attitudes as well as his supposed anti-feminism (see Weiss 1987).

The latter refers to how Rousseau is a rather "immodest witness" when it comes to extending his critique of progress to topics like gender, sex and feminism. Much has been made throughout the centuries of the contradiction between, on the one hand, Rousseau's behavior and personality and, on the other hand, his defense of virtue and solitude. He is anxious around others, paranoid (about Hume for example), he lives away from others but is always surrounded by them, particularly women on whom he depends. And, of course, there are the many descriptions of stereotypical femininity, which put together make it appear as if he singled out women to be educated as "submissive and dependent" (Weiss 1987, 81). Yet, Rousseau is quite unique in his explicit rejection of any basis in nature for the inequality of men and women. It is not because of their nature that he ends up depicting women as educated differently than men. It is inequality, their different social status in society, that sets them apart from men. The unequal situation of the sexes are "the result, rather than the cause of their social role" (ibid, 86) and "natural differences are not very great, inevitable, immutable, or politically relevant" (ibid 94). This position is integral to his basic formula for the state of nature. Women take their place in society in accordance with the inequality that they suffer, and their roles, like those of the scientists, are yet again illustrative of the many personifications that inhabit his work, each illustrating the inevitability of the loss of natural freedom.

Accordingly we can investigate Rousseau along similar lines as how Haraway sought to "queer" Hobbes. Hobbes is its critic, one of the actors in the drama about the scientific revolution, the one who opposed "the founding gesture of the separation of the technical and the political" (Haraway 1997, 24). Also Rousseau, like Hobbes, does not enact such a separation. He remains relentlessly skeptical about the sciences, which, as Cassirer put it, imposes a "false ethical order of things". This is the problem, part of the tragic conception of freedom, because it reduces knowledge "to a mere intellectual refinement" (see Cassirer 1989, 57). This implies that knowledge must claim no absolute primacy in society, and notably includes observations on how man will not change through "historical or ethnological knowledge" (1988, 15).

The latter is particularly relevant when noting how Haraway observes that masculinity is described as given within the context of Boyle's experiments and how gender is remade in the process of science. It is not just exclusion that can be observed in the experimental setting but a process of "crafting gendered way of life". It is not just that witnessing the experimented was a heavily restricted activity, requiring a certain type of modesty. The "general" absence of women began to be established in the experimental life, unlike the engineers involved in building the pump, who were invisible but present (see Haraway 1997, 28). Accordingly this experiment becomes an example of the type of knowledge that seeks "to raise itself above life", and with Rousseau there is still an objection to how the experiment can put its knowledge-claims on the "laws of external things", as if this can be seen in isolation from the tragic circumstances of modern society (Cassirer 1988, 20). In turn, we can revisit the copyright and trademark signs of "FemaleMan© and Oncomouse™". There's still no reason to give up on challenging "the line between subject and object, original and copy, value and valueless". To do so, we are told, is to recognize how both share kinship; they are both "natural obscenities" and s/he show a direction for "a reestablished commons" and "an expansive and inclusive techscientific democracy" (ibid, 71–75).

Also, this is part of Rousseau's program, which includes a critique of property, affirmation of common property and its legitimization by way of a social contract (see Chapter 3). Hence there is a straightforward path that returns us to the "freedom-machine", maybe with a copyleft sign "☺" replacing the copyright sign in the title of Haraway's book. Such a freedom-machine would, thereby, share in this kinship, albeit as yet another type of corporate obscenity (see Tamminen and Deibel 2019, ch 6). What we are aiming for is the ability to re-imagine (as method) and to strategize (as politics) the kinship whereby the commons are being re-established today. What is obscene, however, does not simply refer in a positive sense to the transcendence of dualisms of masculinity and femininity. It also includes the commercialization of the new commons, as shared resources managed within the context of modern capitalism and the class and neo-colonial politics that it implies.

There is a highly selective affirmation of the commons in the context of science and technology, as modern, dynamic and even cultural. The opposite holds as well, that to sideline the commons politically revolves around their identification as traditional, static and natural. Taken together the naive advocacy of the commons is likely to perpetuate the mistaken and persuasive belief that when resources are open to all, they can be equally exploited by all (Sunder 2007, 106). Such naivety is, of course, always possible when building on a critique of the early modern origin of the experiment, which includes replacing Hobbes with Rousseau's tragic view of life in nature, as it still sees freedom as invariably lost and exacerbated by the status of the experiment in the sciences and in modern society. Nevertheless, Rousseau would see such exploitation as abusive and certainly not as a legitimate exercise of sovereign power.

Before discussing this aspect of his work below (his relation to colonialism and imperialism), a queered Rousseau is still sensibly a part of a more methodological approach. His lack of modesty when it comes to gender relations is often noted and he certainly is not an

18th century advocate of "FemaleMan". This is no different than how his critique of modern society did not comprise a defense of the commons and commoners, even if he criticizes modern society. However, he does occasionally flip gender stereotypes around, and therefore the "©" is again an appropriate symbol for what governs that process. One of these moments is worth quoting in full:

> I am far from thinking that the ascendancy which women have obtained over men is an evil in itself. It is a present which nature has made them for the good of mankind. If better directed, it might be productive of as much good, as it is now of evil. We are not sufficiently sensible of what advantage it would be to society to give a better education to that half of our species which governs the other.
> (Rousseau 2003, 19)

This passage might go some way in showing that any accusations of prejudice and bias might be overstated when it comes to Rousseau's philosophical character. It might be a conventional romantic idea that the sexes depend on each other but this interdependency is not given to us as a simple affirmation. It is compared to how great works are neglected while mediocre works are admired in the sense that each of the sexes is taking shape in terms of the other's lack of courage in a corrupt society.

Alongside the quotation that began this chapter – the one about people being mean, corrupt and miserable – our empirical reservations should not overshadow that Rousseau's position is ontological about the existence of social inequality and his ideal-typical commitment to interdependence should be understood accordingly. Also when it comes to the position of women in 18th century society, there is no priority given to any type of attempt to establish the truth about nature. Any empirical observations can be revised relative to our hypothesis about the state of nature. This also applies to gender, which is to say that historical contexts should be changed and can be changed given what we choose to do. However, this will be extremely difficult, and more likely than not we remain trapped in a tragedy.

In this sense it's worth exploring the topic of education, which for Rousseau is not peripheral and has direct bearings on how we understand his views on gender and the relation of power and knowledge. Consider that Rousseau became a celebrated novelist after writing about a young woman and her tutor in a society that cannot accept their love (Julie, ou la nouvelle Héloïse 1761, see Rousseau 1997). A year later he published Émile, a political treatise and educational fantasy that subsequently turned into one of the founding works on pedagogy (Rousseau 1993). Both Julie and Emile are obviously fictional, and such fictionalization is not accidental; both main characters illustrate a transcendence of our tragic circumstances that has as a key component how we raise our children, educate them into citizens, or otherwise teach ourselves to internalize a conception of freedom that requires us to live in accordance with duties to the wider community. On the one hand, Julie and Emile experience freedom in society as personification of what Hannah Arendt describes as "the new sphere of intimacy" (Arendt 1997, 39). They are personifications of the expansion and enrichment of individuality that has come to be seen as integral to the commitment of liberal democracies to the freedom and equality of individual citizens. On the other hand, Rousseau has them learn to accept the limits set by what he sees as the necessities of life, which controversially includes the presence of authorities and the need for conformism when living in a community (Ohana 2017).

Accordingly Julie experiences romantic freedom but comes to accept having to marry another. Rousseau (in book V) writes about how Emile finds his ideal wife; she is called Sophie and she is cared for by her mother who plays a role similar to the one played by Emile's tutor. To become a model for a social pact freely associating individuals, which includes their love and their marriage, is only possible after having been subjected to the abstract pedagogical principles of their tutors. The resolution of such a clash between the freedom of youth and the educational project mirrors Rousseau's state of nature theory. Especially Emile's education is yet again a journey out of a state of nature. He has to give up the natural freedom out of self-preservation. As he grows older, childhood is invariably lost and "to try to remain in it when it is no longer practicable, would really be to leave it, for self-preservation is nature's first law" (Rousseau 1993, 187). Emile, in this sense, grows up in ways that extend the formula of Rousseau's state of nature theory: natural man is alone, lives in solitude and is self-sufficient. Such freedom, however, is to be left behind and he has to be become educated so that he can establish relations with others. And, once more this is clearly not intended as an empirical or realist position, just like Emile's upbringing was not meant as "truth".

The following quote articulates this with an interesting reference to scientific practice:

> Let us begin then by laying facts aside, as they do not affect the question. The investigations we may enter into, in treating this subject, must not be considered as historical truths, but only as

mere conditional and hypothetical reasonings, rather calculated to explain the nature of things, than to ascertain their actual origin; just like the hypotheses which our physicists daily form respecting the formation of the world.

(Rousseau 2003, 50, 51)

We were never reading a series of logically consistent empirical statements about educational strategy. Rousseau's fictions and hypotheses are part of a method, like that of "our physicists". Applied to Emile it implies that once he grows up he can be considered as the modernist ideal of the new man who might choose to be free. However, the method establishes a relation between fact and practice that revolves around Emile learning to recognize that conformity is a necessity.

Such a formula for the transcendence of nature is one that practically attempts to orient our education towards the ideal of a new man whose freedom is modeled on the many precarious relations that can conditionally help to support the nature of things. It does not matter that whatever counts as "necessity" might be difficult to establish in exact terms. The child grows up regardless. However, the conditions that characterize the journey of a child out of nature do matter. There are many inequalities in society that have to be avoided temporarily, during his education. Meanwhile Emile, like Sophie and Julie, is being tutored in such a way that he learns how to deal with the conformism imposed by society. The goal is, therefore, to learn how to accommodate the opinions of others, which includes forms of cooperation (like the family, like work etc.) and other new ways of providing for our needs that create necessities that chain us. In this sense their inevitable journey out of nature includes acts of altruism, cooperation and the fruits of living and working together as well as selfishness, fear, misery and insecurity.

Included in his education is how Emile learns that study does not lead to truth and that an education wherein knowledge is a decoration and does not bring happiness. Such lessons, however, also imply that he is taught to accept authority, he must serve, conform to others and so forth. At its furthest extent it is even the case that Emile can be said to be conditioned into a person who chooses what his teacher taught him. This overall lesson is that he must choose to bind himself to the common benefit because in the encounter with others it becomes impractical to remain outside of political society. Does this imply that Emile's (and also Julie's) freedom is an illusion? After all, on a realist reading he has been absorbed into a comprehensive pedagogical project wherein his tutor has total control. The formula suggest that the teacher is the authoritarian personality that acts as the embodiment of the unifying logic of popular sovereignty. If the goal is to free Emile step-by-step to face a world that has to be remade constantly, how can such an education be expected to deliver the type of independent-minded individual whose civic engagement actualizes freedom practically within the confines of the political life of a community?

Only by acknowledging that Emile is not a realist position are we left with anything else than a model for conformity as an inescapable characteristic of the human condition. Otherwise Emile turns into a victim of a romantic educational doctrine or a model for the comprehensive re-education ideologies and indoctrination that have been inflicted on populations across the world. Included would be a realist understanding of Rousseau's seeming commitment to separate educational models for men and women, which continues to have its appeal wherever conservative types of leadership are the norm. Furthermore we would have to extend such an oppressive logic to science in society and the attempts at its democratization. On this premise STS might have a point in dismissing its own roots in political theory, given where it leads. Or, instead, we concede willingly that it is not realistic to assume that man exists alone outside of society in perfect solitude, and that Emile, Julie and Sophie are fictional. Both are means to an end, idealizations of the precarious relations to nature that Rousseau mobilized in his rebellion against the conformism inherent in modern society (Arendt 1997, 39–45).

Nature, science and colonialism

Setting apart Rousseau's appeal to our imagination from Hobbes' commitment to realism is important when considering topics like conformism and education and the status of gender in an unequal society. This includes how colonial mindsets still have their hold over the present and how idealizations of nature were integral to how colonialism operated.

First, Rousseau affirms that disorder and insecurity characterize men's relationships to each other. Man's journey out of nature matches that of Hobbes to a remarkable extent, with as the difference that the violence is a result of interaction and man's sociability. Rousseau's formula affirms a position of initial equality – a condition wherein no individual can hold a natural right over another because they have not yet encountered others. His view of what counts as a legitimate exercise of power changes accordingly – leading him to seek liberty for all citizens in a civil state and evaluate the role of the sciences from this perspective.

This is again the major theoretical difference with Hobbes. For example, in Leviathan it is mentioned at various times how countries at all time exist in a state of nature (ch. XIII and XXX) and this includes observations that establish that violence characterizes life of "savages". This is not a mythical past, but a present-day reality to Hobbes. Natural men exist everywhere outside of civilization in Hobbes work, most notably in the Americas. Their life in nature is treated as an empirical reality, observable everywhere else and disorder is a natural condition. The more disorder is observable, the more it is proven to be a natural foundation that supports a view that order must be established and maintained at all cost (see Tully 1995).

Furthermore Hobbes's formula is only one of many examples. Grotius, Locke and many others similarly described international order in terms of such realist observations (see Tuck 1999). These accounts might differ in their details but there is invariably an essential disorder that has been naturalized with only some civilized societies who, along with citizens that turn into colonizers, are surrounded by savages (Tully 1995). This is not simply an outdated position, a colonial shadow that has been gotten rid of, long ago. It can quite easily be shown that it retains its hold over today's "realism" in international relations (see Huntington 2002, Kaplan 2002). Consider how much of the news shows us life outside of the societies, selectively making empirical observations to demonstrate that the savages are coming, for your land, your property or your life. The formula remains the same, based on this type of "realism" the conclusion is reached that civil rights are luxuries and your freedom should be given up in favor of security in a world otherwise characterized by famine, war and any other type of misery and disorder.

Second, Rousseau might not have been directly involved in colonialism (like Locke and Grotius), his work is still full of statements about the independence of natural men. Such views have operated as a foil for the exploitation of whomever would sooner or later find themselves caught up in the expansion of the European world. Rousseau's era was characterized by the expansion of a global network wherein success revolve around how new agents, artifacts and distant peoples were linked to dedicated spaces, such as plantations, model farms and botanical gardens. Such networks became entrenched as patterns of domination, dispossession and exploitation that were deeply invested in the sciences and societies of Europe (Muthu 2003, Parry 2004, Kloppenburg 2005).

Rousseau was well aware of the prejudices that went into the travel accounts about the rest of thew world. In "a discourse on the origin of inequality" he states that he seeks to avoid "the blunder made by those who, in reasoning on the state of nature, always import into it ideas gathered in a state of society" (Rousseau 2003, 65). Effectively this implies that he affirms "the universality of sentiments inherent in human nature" while shining "a spotlight on the forms of abuse and oppression that had originated as a result of denying those universal sentiments" (Gozzi and Valente 2021, 45). Accordingly science, art, social relations, even language are not to be naturalized, as characteristics of natural man. The state of solitude implies that man is alone and it ends immediately, as soon as interaction is established and relations begins to multiply. There's a clear separation of nature and society, as a theoretical foundation, which states unconditionally that men are naturally equal and free. And there's an instant critique, of how natural freedom is instantly subjected to the corrupting influences of anything cultural, and has no place among the thoroughly artificial character of modern society.

The end-result is a situation wherein there are many seemingly realist observations, which includes many that are about non-Europeans as savages, and yet these are secondary to the affirmation of the hypothetical, fictional and idealized character of the state of nature.[5] What this implies is that Rousseau does not believe that non-Europeans are free from the vices of modern societies, after all, these are introduced the moment that men become social creatures. Yet, non-Europeans are still frequently compared positively to the corruption of European societies. He discusses Caribs, Hottentots, Amerindians in support of his original state of nature but also to indicate the inevitable loss of this state. For example he identifies the state of nature to life in the Americas, because there is no large scale agriculture and metallurgy (Muthlu 2003, 34–39, see also Rousseau 2003, 92).[6] Muthlu describes the ambiguity of this "middle stage" of the journey out of nature and into a state of society. Rousseau was quite open about turning this middle stage into a golden age alongside the virtues of ancient civilizations, both were key components of his critique of progress. However, he does recognize that violence and vice are characteristic of this stage, even if it still lacks the "egregious injustices of the civilized stage that Europe is in" (ibid).

The problem is most obvious when Rousseau is giving examples to show that people in this stage remain closer to a state of nature. This leads to sometimes disturbing discussions. Rousseau's belief that meat eating is not natural, that certain physical skills were more developed and that orangutans might be closely related to humans might still hold some currency today. These examples, however, are given by

Muthlu along with observations about the limited cognitive abilities of certain non-Europeans and their similarities with feral children (ibid, 40–42). Consequently it can simply be affirmed that this stage-theory of human development in realist terms does not see human being as social and cultural beings under all conditions. This is highly problematic even if he, unconditionally, affirms a shared humanity.

Consequently he keeps nature and society apart, even if he holds a negative view of the latter and without the intention to dehumanize any "savages". This is a given in the works of Grotius, Hobbes and Locke, which continue to have their hold today over what type of "individual" ends up being rewarded "rights" and what type of "people" gets to be "sovereign" (Tully 1995). Rousseau, however, had little interest in "national self-assertion" and even his appeals to patriotism are about self-sufficiency and avoiding the conflicts that arise from dependencies on other nations (see Dent 2005, 46). This includes trade, which "serves to complicate and extend the ties of international dependency and envy that are the cause of war" (Roosevelt 1990, 99). In his advice to Corsica and Poland he goes as far as recommending to disengage "the nation's character from foreign influence", a type of "economic autarky" (Hoffman and Fidler 1991, xxxiv, xxxv). The ambition to be free does not, however, correspond to any type of nationalism that would turn expansive; whomever wants to be free "must refrain from becoming a conqueror" (ibid, lxi).

On the one hand, Rousseau is quite undervalued for his premonition about how countries would soon be based on "national sentiment", how commerce does not "breed peace" and he even shows little faith in peace as guaranteed by international norms and the role of international organizations (ibid, liii). Only "islands of peace" and "moments of grace" are attainable (ibid, lv). These exist among conditions that are aimed at perpetuating the inequality that is characteristic of a state of war between countries and it is in this sense quite clear that Rousseau thought colonial empires were illustrations of the corruption of modern society. On the other hand, it's also the case that (unlike Diderot) there's no fully formed, comprehensive critique of colonialism or defense of peoples he held up as more virtuous against it (ibid, 46). This critique holds even though many of those types of political philosophies were inspired by Rousseau. Muthlu describes Rousseau's position (along with Diderot) as "a multidimensional social theory" that "recognizes the complex interdependence of structural and voluntary features of human life" (ibid, 46).

Accordingly we again should remind ourselves of the "method that informs his speculative history" (ibid, 34). Rousseau sought to distance his intellectual framework from the realism of his predecessors. He was well aware that his empirical observations might be inaccurate, and his references to nature and society, as basic categories of natural philosophy, did not require that they were established decisively or deterministically. The conclusion, therefore, is a potentially damning contrast. We criticize Rousseau for not theorizing human nature as social or cultural, and this extends from indigenous peoples to femininity. What, then, do we do with today's attempts to engage with the multiplicity of relations to nature, empirically, to characterize contemporary society in its many relations to science and technology?

After all, post-Latour we've let "nature" back into our analysis, restricting the role of social categories and sociological "truths". This position implies that there's no way of severing our understanding of what a human is, as an individual, from the type of naturalism that comes with its understanding of science and our close ties with our technological devices. This type of hybridity is seen as omnipresent – as the back and forth between science as nature and social order – and even characteristic of being human, as the cyborgs or posthumans we already are. And if such critical naturalism characterizes a considerable part of STS's conceptual lens for our contemporary situation, as practice and in terms of performativity, what could be the reason that the same critique we apply to Rousseau does not extend to ourselves?

Conclusion

In all likelihood those who only know of the doctrine of natural philosophy by hearsay, or look at it through the lens of 19th- and 20th-century philosophers whose sensibilities are closer to our own, will simply see a teleological doctrine, Rousseau included. After all, also he seeks a restoration of nature, which therefore must have some essence, soul, spirit or consciousness. This chapter, however, has sought to show that his thought should be appreciated has having meaning beyond its immediate historical context as it is – to a reasonable extent and potentially – consistent in its method and, by extension, in its usage of core concepts that inform ours, politically and in terms of modern research methods.

If it has not been enough to do this in the form of exposition and argument, consider again how Rousseau could still qualify as an example of doing "case studies" if we are willing to consider fiction in its relation to the empirical world, either his novel or the life of Emile. And if that's not enough he was an accomplished participant in the culture of science and "nearest to being an anthropologist" (see Levi-Strauss 1961, 389, Luhrmann 1990).[7] Of course, the affirmation of these parts of Rousseau's work are not going to make much of a difference. If we criticize Rousseau, as a good anti-foundationalist would, we still dismiss his method, as well as his hold over the theoretical foundations of STS. Accordingly we're left with the widespread fascination with the process of naturalization, in STS and adjacent fields, and how reality is full of the anthropomorphization of machines, the blurring of the lines between "humans" and "things" and the plethora of metaphysical and ontological discussions this opens up.

Like a shadow, foundationalism travels with anti-foundationalism, it follows epistemological claims, attached to the very best cases or illustrations, to sophisticated reflexive language turning into concepts and theories that attach all the right disclaimers to celebrations of hybridity and multiplicity. Rousseau travels along with us when we are busy with our "revolt against Mertonian sociology" (Collin 2011, IX)? He nods approvingly to those who seek to carefully describe nature, its many precarious relations and avoiding any type of essentialism in its empirical description. Some might make "strong" claims about the social origins of science with the sociologist as its "ideal practical reasoner" or a "weaker" one wherein social reality simply determines what knowledge ends up useful (Collin 2011, 59). Both are simply discussions of the state of nature and its status does not hinge on its realism, how exactly our empirical studies of science and technology establish a claim on social reality.

Certainly they are integral to the method, as means whereby to reach a perspective whereby to establish the type of "science in society" to aim for. He is highly critical of the social status of science but there might be cases wherein it's desirable to temporarily ascribe "realist" status to certain types of expertise, with the associated skills existing "independently of public recognition" (ibid, Collins 2004). This is, again, a means to an end; it is useful to have an inventory of cases that show how different types of expertise and the associated skills lay their claim on nature and how they operate within a wider society. Yet, this is not the goal, given the status that empirical work on nature and how each case invokes a relation to nature that, invariably, reflects a type of foundationalism. Don't be surprised if, just as with Rousseau, foundations turn up uninvited, as happens constantly with the many attempts to affirm distant peoples, their local circumstances, their intricacies and their values. There will be something of Rousseau following you around, as a shadow that turns out to be, rather, like a state of nature we have to leave, like the social contract theory that matches it.

In sum: deconstruct away, sideline Critique all you want, be proud of the scientific reality you fought so hard to get access to all you want, just don't exaggerate your position on empiricism to the point of mysticism. Anything comes with historical foundations, whether or not we affirm these (as intrinsic or with some type of dignity) or deny them (as a social and historical construct). We don't have to commit to any type of outside, as real, as mythic, as nostalgic but neither does deconstruction make them disappear, ontologically they are there, with their ambiguous universals, in how new hybrid beings come with new subjectivities that turn into the next (ontological) constitutions. Let's not lose ourselves into this infinite regression, flatten our observations because it is what the methods of the field tell us to.[8]

Its implication would be a claim on society and its betterment that is solidly within the confines of political theory, particularly the suggestions about how to better situate science in society. It's a tautological type of sociology, still trapped in its own terminology, but this time through its conviction that there should be no ontological commitments to any categories at all. STS has been raising the stakes for a while now, extending its empiricism beyond the older agendas of overcoming the philosophical preconceptions that characterize the inner worlds of science and technology. Going by the success of Latour, STS is now visibly political and normative about its ontological monism; it's "the future of mankind" that is at stake, even if that agenda was there all along (see Collin 2011, 131).

Indeed let's be political about science and technology, but let's be straightforward about the ontological commitments we put down. In this book it is the loss of freedom in the context of software and our critique and methodology are derived from Rousseau. Earlier it was Foucault and sovereignty, extended to Rousseau and the history of crop diversity, GMOs and modern agriculture (see Tamminen and Deibel 2019, ch 5) while Grotius's law of the seas extends to marine biology and Marx's species-being applies to synthetic biology (ch. 7). Maybe in the future it might be interesting to go directly after the historical roots in political theory of today's empiricists. There's Locke and Hume for example, and how they relate to Dewey's status in STS. The point, however, would be the same, which is to mobilize their paradoxes and contradictions when it comes to the suggestion that there can be anti-foundationalism in empirical studies of science and technology without clearly defined ontological commitments.

The premise, of course, is that the political philosophical roots of studies of science and technology require that such "cases" are to be interpreted as a group and constantly expanded, whether is through the engagement with early modern, modern or post-modern theories. This chapter proposed taking Rousseau seriously as part of a general way forward, methodologically and alongside other such steps. His method is no exception in how it moves on – from deconstruction (this chapter); after being weighed down by such "logical chains" what follows is, of course, a "flight forwards" into more discrepancies (Althusser 1988, 115) and there are always "supplements" (Derrida 1988). Let's invite them in, but only as means to an end, and with Rousseau that end is not reflexivity for its own sake, that sidelines our experience of tragedy along with any meaningful relations to political theory. This is next: turning to the tangible implications of understanding his conception of "freedom" methodologically, which is to say as a meaningful category in times when the intellectual and material emptying out the concept of democracy in the context of science and technology appears as a given.

Notes

1 I'm referring to the debate between the Edinburgh School and the strong program with Latour. The former adopted their symmetry principle only for the explanation of science and the understanding of nature that this implies, the latter extends it to include much more, comprising whatever exists (see Collin 2011).

2 Bernardi and Bensaud-Vincent explain in detail how Rousseau behaved "like a typical enlightened amateur, reading scientific literature, conducting laboratory experiments and field work, discussing with experts, teaching, and making his own books and collections" (2012, 68). Furthermore they mention that that Rousseau supported Linnaeus over Buffon, popularizing his thought in France. As indication of the relevance of this dimension of his writing: Goethe was still reading Rousseau's observations on botany, even seeing himself as following in Rousseau's trajectory (see Hammer 2005). In turn, this emphasis on Linnaeus would open up an interesting perspective on how it is occasionally presented as straightforward that Rousseau took Buffon's essential distinction between animals and humanity and applied it to the distinctions between savage and civilized man (see Pagden 2013, 157).

3 As was mentioned in the introduction, Popper described Rousseau as a dangerous romantic collectivist, but this is a reference to the expansive role of the state (Popper 2002, 50).

4 See Ohana for a detailed discussion of how Rousseau interprets the story of Prometheus. It is worth mentioning, however, how Rousseau explained to a "reader" (in reality Claude-Nicolas Lecat) the role he sees for Prometheus. He wrote: "I would have thought I insulted my readers and treated them as children if I interpreted for them such a clear allegory; if I told them that the torch of Prometheus is that of the sciences, made to animate the great geniuses; that the Satyr, who seeing the fire for the first time, runs to it and wants to embrace it, represents the vulgar men who, seduced by the brilliance of letters, indiscreetly give themselves over to study; that the Prometheus who cries and warns them of the danger is the Citizen of Geneva" (see Ohana 2017).

5 Another one of the passages that gives a sense of Rousseau's description of the state of nature is this one from the discourse on inequality (2003, 80): "If I have expatiated at such length on this supposed primordial state, it is because I had so many ancients errors and inveterate prejudices to eradicate, and therefore though it incumbent on me to dig down to their very root, and show, by means of a true picture of the state of nature, how far even the natural inequalities of mankind are from having that reality and influence which modern writers suppose". This should already go some way to establish necessary context for how we should treat Rousseau's studies whenever realist statements are encountered.

6 The passage that is being referred to is worth reading and goes as follows (2003, 92): "Metallurgy and agriculture were the two arts which produced this great revolution. The poets tell us it was gold and silver, but, for the philosophers it as iron and corn, which first civilized men, and ruined humanity. Thus, both were unknown to the savages of America, who for that reason are still savage (...) One of the best reasons, perhaps, why Europe has been, if not longer, at least more constantly and highly civilized than the rest of the world, is that it is at once the most abundant in iron and the most fertile in corn".

7 Levi Strauss doesn't hold back: "Rousseau, our master and brother, to whom we have behaved with such ingratitude but to whom every page of this book could have been dedicated, had the homage been worthy of his great memory" (Levi Strauss 1961, 389).

8 It might appear to some as if these passages are contradictory given my work on Foucault. This would be correct were I to agree to characterize Foucault as an "anti-humanist", referring to how he discussed "modern man" simply as a historical category, a historical formation, one that has been understood through the anthropomorphism that is implied in humanism, as an end in itself. At its extreme the implication is that subject can be declared dead and man has begun fading out of history along with the categories that derive their meaning from the subject. This is not helped by how "life" continues to be relocated to the domain of the (life) sciences while, historically, and philosophically it belonged to the realm of the transcendental or divine. My approach, however, has been to show Foucault's anti-Hobbesianism, what he called "the problem of sovereignty". Accordingly we find ourselves once again with the main

characters of the problem of sovereignty, the early modern of doctrine of natural law and I include Grotius, Locke, Hobbes, Rousseau, Wollstonecraft and so on (see Tamminen and Deibel 2019). Their respective naturalisms, at times naive and sometimes sophisticated in their hybridity, are therefore part of the method whereby to approach the centrality of the subject in modernity and any such suggested endings. Accordingly Foucault addressed specific configurations of the subject and its end. He does not thereby finish off "man" and every single type of commitment to "freedom" or any other such modernism. These remain possible even if we recognize that certain conceptions that are of particularly recent dates might "be erased, like a face drawn in sand at the edge of the sea" (Foucault 2002, 422).

3 Social Contracting and Freedom

> There are two kinds of dependence: dependence on things, which is the work of nature and dependence on men, which is the work of society.
>
> Rousseau, Emile (1993, 58)

Introduction

The quote above might seem, at first glance, to echo the familiar assumptions about a split between technology and people. Technology is, in this view, a means to materialize nature into things, alongside science as the means to establish privileged relations to nature. This would, however, lead to the wrong conclusion about Rousseau's methodology.

If this quote had been about the separation of technological things and the social world of people, the appropriate response would likely have been to examine such a demarcation, critically or methodologically referring to "boundary work", "boundary crossings" or similar type of approaches that seek to characterize the inevitable back and forth across the line (see Gieryn 1983; Halffman 2003, see also Tamminen and Deibel 2019, Chapter 4). The phrase could, thereby, be understood with more epistemological nuance and precision, a division of things and people along similar lines as those of the frontiers of science, the limits to growth, the transgression of biological limits or other such types of (ontological) boundaries that exist, are maintained, challenged, promoted, materialized, naturalized and so on.

However, "the dependence of things" is not a reference to the materialization of things, as the outcome of science and technology and as opposed to society, culture and politics. Instead, it refers to what Rousseau calls "necessity", or even "the law of necessity" and its meaning is simple enough when approached in terms of the

DOI: 10.4324/9781003193609-4

education of Emile. Rousseau explains repeatedly that a child desires only what it needs, as he does in the next paragraph, which begins with the exclamation: "keep the child depending on things only" (Rousseau 1993, 58). Accordingly, the book is organized around the stages of development that Emile goes through before reaching adulthood, but at the same time every phase is explained with constant references to political theory. Rousseau morphs pedagogy into politics, mixing nature and culture into an integral part of his social contract theory.

On the one hand, Emile as a young child should not be introduced to novelty, because each new "habit adds a fresh need to those of nature" (ibid, 34). Echoing contemporary debates on over-protective parenting, the chapter started by stating "let it cry in vain" (ibid, 49, 50) and later the advice is to let a child "jump, run and yell" (ibid, 58). On the other hand, this type of educational necessity is intended to "leave the child the use of his natural liberty" (ibid, 62). Emile, or anyone else of similar age, is "neither beast nor man, but a child" (ibid, 57), even though, invariably, this type of dependence on things will end. And that means tragedy resumes; after all it means the exposure to modern life and its corruption ends up turning those that are free into "slaves to the community" (ibid, 62).

Consequently things and people should be kept apart, maintaining the dependency of things as long as possible and letting the law of necessity take its course. Accordingly the split between these types of things and people should be maintained, the demarcation is intentional and its preservation has as its purpose to give the child an "education of things". It is a delaying tactic, as the advice is to prevent jumping "from the objects of sense to objects of thought" (ibid, 156). Eventually the moment will inevitably arrive, as a necessity, that Emile is introduced to society, albeit with a qualification: there is an "*if*", the word that introduces the next part of the citation that started this chapter. While difficult, it is worth reading in full,

> *If* there's any cure for this social evil, it is to be found in the substitution of law for the individual; in arming the general will with real strength beyond the power of any individual will. *If* the law of nations, like the laws of nature, could ever be broken by any human power, dependence on men would become dependence on things; all the advantages of a state of nature would be combined with all the advantages of social life in the commonwealth. (...)
> *Keep the child depending on things only.*
>   (Rousseau 1993, 58, cursive and underline added)

The "if" is a reminder, as discussed in Chapter 2, that it is not decisive whether or not Emile's formation into a member of the community has been accurately depicted in empirical terms. What matters is the focus on the goal of his development, as a thought experiment aimed at freedom in society. This combines education, the social contract and the geopolitics of the law of nations, each of which is presented in a relationship to a state of nature theory that is hypothetical in each case.

What is not hypothetical, however, is the basic formula: that the general will ("Law") comes in the place of individuals, each with their own particular will. After all, this is the way out, and it revolves around "Law", or rather around Rousseau's famous concept, which at its most general asserts a condition of basic equality between the associates. The formula is not difficult in how it acknowledges that equality "is the condition for the freedom of each"; each member who joins a political association is granted humanity and moral dignity by everyone else and, reciprocally, grants it back (see Dent 2005, 152). Simply put, Rousseau seeks to establish that there could be a different type of freedom, one that is guaranteed by all to each other within the political association, as a legitimate social contract with sufficient appeal for everyone to have joined voluntarily. Of course, this is still a type of escape – from the various ways wherein natural freedom is being lost, and its details require more intricate description. How, for example, might this "arming" of the general will be imagined? And what about the promise that dependency on men could turn into dependency on things?

It is as a method (the "if" again) that this chapter will engage with the constitutionalist offshoots of the Latourian vernacular. What this implies is that there are many types of "nature-cultures" in the quote above, each of its dimensions invokes "quasi-subject" and "quasi-objects" or any other similar language to affirm hybridity in between nature and society. Each of them is still characterized by some degree of skepticism about dualism, origin stories and other types of foundationalism but approached methodologically. Our goal is to take our analysis beyond simple affirmations that there are many such hybrids and that everything is at stake in how nature and culture mix.

Rousseau's method has as its objective to escape the tragedy, to get rid of what it calls "social evil", which refers to the corrupting dependency on men. Such hybridity signals at a tragedy, a type of social decadence but one that is not entirely inevitable, in both its structuralist and constructivist senses. Indeed, it "ought not to be borne", by anyone; this is what the citation refers to as the goal of "law", which at the time was seen as breaking of the laws of nature. Its implication is simple; it takes the analysis from how man is born free, as stated in the first line of "the social contract", to its full implication: that the legitimacy of a political association depends on its rejection of anything less (Cassirer 1988, 29).

The method and the contract

The "if" signals at a method that is ontological in the sense that "we should "not forget that in our day we must be superior to this law" (Rousseau 1993, 58). Accordingly, Rousseau's tragic conception of freedom has as running towards our chains, giving up our claims on our freedom, but this is not the part that is abstract, controversial or impractical. Quite the opposite, his method is easily understood as fully compatible to the anti-foundationalist and anti-essentialist stances that characterize many of today's social methods whereby science and technology are understood (as discussed in Chapter 2). It is the goal of the method to deliver its ontological claim, the necessity, desirability and practicality of freedom as a meaningful category in society.

Certainly this was something that Rousseau addressed, right from the start of Emile, in its preface, he states:

> people are always telling me to make practicable suggestions. You might as well tell me to suggest what people are doing already, or at least to suggest improvements which may be incorporated with the wrong methods at present in use.
>
> (Rousseau 1993, 2)

These "wrong" methods are everywhere. This book started, in Chapter 1, with how Rousseau is still mobilized to affirm a type of altruism and as naive in his advocacy of a return to nature, or a lost paradise wherein man is still "good" or "authentic". If he has such a position, it at most implies linguistic confusion surrounding his affirmation of passive individuals who in solitude and isolation do not yet know about good or evil, selfishness and altruism etc. This means that there remains potential for these attitudes and all of them exist, even if we acknowledge that none of them is given at the level of nature. Accordingly, their presence has to be acknowledged, but always within an overarching methodological objective: to establish a framework that seeks to cultivate an open-ended relationship that ranges from natural freedom (the dependency on things) to a position from where there is at least some potential to affirm "freedom" as it could exist within a political association (dependence on people).

This applies directly to our own relationship to the various types of nature-cultures, to hybridity, if we acknowledge any type of active legacy involving Rousseau. Recalling his proximity to Hobbes, as discussed in the previous two chapters, implies that our own understanding of today's predicament is subject to its "most frightening" aspect: the state of war as a perpetual state, as a description of the state of society, extended to any aspiration of Enlightenment (Althusser 1988, 100). Rousseau's formula takes over Hobbes characterization of nature and turns it into a social condition. This stretches any strict legalistic notions of "the state of war" as the opposite of peace, with animosity, hostility and the fearful character of men that Hobbes foregrounds becoming a normal part of many modern societies that are not actively engaged in acts of war (Roosevelt 1990, 56). The end-result is that Rousseau characterizes the journey out of nature as gradually leading to dependency on men and thereby turns into a state of war, effectively only postponing what Hobbes started with.

The difference, however, is that we do not aim to end up with something akin to a Leviathan. This could either refer to any type of dictatorial sovereign with full control over a country's security or the considerable "emotional force" that comes into play whenever an authority figure invokes legitimacy as derived from a type of paternalism, as a "father figure" or "father of the people" (Dent 2005, 127). Within the context of Rousseau's method, these scenarios are both examples of the naive naturalism, of the "wrong method" at work. The same, of course, applies to the type of romantic escapism that Rousseau is popularly (mis-)identified with. The latter is just as much everywhere in our own popular culture – in western museums of primitive art, films on exotic untouched nature-aliens on distant planets or misunderstood natives who require protections because they are not yet modern like those considering themselves as more civilized.

What if we, at last, retired this tired formula as Rousseau tried to? We can console ourselves that his method allows for mistakes, they are fictions with consequences and therefore we can retain some of our attachment to them, even if other aspects are suspect. What we will have to do is to make them secondary within the wider context of our method. Effectively we will end up doing what Rousseau is infamous for: treating our favorite fictions as part of the contradictions and paradoxes that he is typically blamed for, even though they are usually the ones that still retain their hold over the crisis of the present. Such naive naturalisms are our own contradictions, characteristic of our cultural experience of modernity and from this perspective (his method) we can face much more directly our own need for new ideas about social contracts and about democracy as integral to our own attempts at a critique of progress.

Social contracting, in between Rousseau and STS

Social contracting in STS is a well-established practice, with as its best-known example Latour's non-modern constitution. Also, this proposal was intentionally hypothetical, based on a speculative origin story about the separation of science and society and this is what it remains. After all, Latour and Woolgar conceded that their account is "fiction" in the sense of going on "an anthropological journey to visit the strange tribe of neurochemists" (Collin 2011, 111).

Accordingly an inescapable modernism surrounds the type of "ontological constructivism", which 1) re-introduces a transformed nature in its relation to the material world and 2) turns reality into a construct alongside our knowledge about it, including our cognitive state (ibid, 122). The approach is aimed at overcoming historical determinism (of nature, of progress, of the moderns and their reality). This remains the case when defining the relation to social contracting empirically, as is the case with "bioconstitutionalism". This type of approach still fits within the space carved out by the Latourian call to compare nature-cultures, it is, however, defined methodologically as strictly related to an empirical field. This could be anything, but in this case the concept refers to how a constitution is dynamically and symmetrically caught up in "the simultaneous and mutually constitutive moves in science and law" (Jasanoff 2011, Hurlbut et al. 2020, 983, Jasanoff 2011).

The end-result is, therefore, that there are two closely related methodologies. On the one hand, there is a method that starts with a historically defined problem, conceptually conceived of as a split in between politics and the way science and technology are organized. This demarcation never "really" existed as a natural object and a theory of reality, but it is significant as an enduring fiction that has effects on today's world. Latour describes it as "the entire modern paradox" that captures the basic historical relation that should inform contemporary studies of science and technology. It is "the secret of the modern world" how the exercise of power is derived from the modern constitution through the constant mixing of nature and culture, the countless hybrids that make such constitutionalism possible (Latour 1993, 30). On the other hand, there is a method that does not rely on any such historical theorizing; after all, all such speculation on our modernity (or lack thereof) can be ignored, we should focus strictly on the proliferation of nature-cultures today.

Considering the same case, we could aim to compare "Bioconstitutional Cultures" (Hurlbut et al. 2020, 989), which methodologically seeks to address a split between the world of science and the world of law. That there is such a split is a strictly empirical matter, and the end-result is the same anyway: an attempt to establish similarity and difference from out of the maelstrom of imaginations and practices dealing with the political subject and the legitimacy of procedures. These are perceived as at once constitutional and "bioconstitutional commitments" (ibid). Its ambition is therefore to be simultaneously comprehensive in its application as a method and yet it remains (more) detached from conventional social contract theory. After all, there are limits to the application of social contract theory to our own predicament. Haraway notes, as much when discussing Latour's constitution; stating that our situation is "vastly different from the constitutional arrangement that established the separations of nature and society" (Haraway 1997, 43). In other words, historical continuity is exaggerated when it comes to the nature-cultures that are the decendents of the air pump and how Hobbes and Boyle are at the origins of Latour's critique of the modern constitution.

This makes sense when the application of a social contract theory refers to the world of science and technology, which would simply add to the already considerable complications when it comes to the limitations of conventional territorial constitutionalism. The overwhelming hybridity of facts and objects would thereby come to characterize how the legal constitutions of countries as they exist in the world already have their limits. Specifically it's proven hard to establish meaningful conditions that all would "freely consent to" when seeking to accommodate the recognition of demands from diverse cultures and experiences of identity and personhood (Tully 1995, 7). This gets much harder when we also include bioconstitutionalism among those demands. Obviously, the same applies to Rousseau's social contract. The role of science in society are part of his social contract theory, which is the foremost historical instance of a social contract based on the consent of all and in that capacity has become tied to the "project for a future possible state" (Pagden 2013, 162). This applies to Rousseau specifically, as it is his social contract that is usually seen as tied to the rise of nationalism: as "an anticipation of the Nation as a totality" (Althusser 1988, 98).

In this version of the theory, Rousseau appears as having initiated the Nation as an empirical reality, through his advocacy of civic education, civic religion, economic reform and so on. Each of these dimensions of his work indicates to a state that has "its first aim" to create "the sort of subjects to whom it can address its call" (see Cassirer 1988, 23). While there is an obvious idealism involved in the creation of such enlightened citizens, Rousseau thereby placed "responsibility at a point where no one before him had looked for it". While he condemned society like none other, it's s him who placed "the burden of responsibility" on what society could become (ibid, 32). And this, in turn, is the position for our own attempts to act responsibly when it comes to the split between science and society, whether it is presented theoretically, historically or as a matter of empirical studies.

Let's not be modern/or the legacy of the general will

It is Rousseau's "general will" that Hannah Arendt points to as the decisive historical turning point that gave rise to the mythological frame of societies as "national households", which is to say that "the people" started being considered in terms of a shared interest, with all of its members beginning to act as if they belong to a single household. These nations became the vehicles for authoritarian leaders, as the personification of the interest of the masses. Crucially, this observation on the enduring influence of *oikeiosis*, as a transformation that is traced back to stoicism, is also directly about Rousseau's critique of the sciences (Arendt 1997, 39–44, see also Brooke 2012).

Today's mass societies might show a liberal commitment to freedom and equality or they might not. They might be different in how the few rule in name of the many, but also a new type of mass society and its management of shared interest requires being "at peace with the scientific outlook inherent in its very existence" (ibid). Such an outlook, therefore, extends equally to an authoritarian mass society and to societies that dream of social harmony organized around any type of omnicompetent executive power. Both are examples of a monologic of sovereignty, as contained by the proximity of Hobbes and Rousseau, and, by extension, comfortably comprising the conventional framing of modern science-society relations, public understanding of science, direct involvement in science and technology and other democratic aspirations of STS. Such a theoretical horizon does not refer solely to the modern state. As Arendt explains, the harmony of self-interests has also been superimposed on the economics of modern society. The individualistic fiction of man, as isolated, and as a strictly self-interested creature has long since turned "law-like", whether it is the context of a communist utopia or of the "invisible hand of the market" (Arendt 1997, 42). Ironically, liberal and socialist economics are each equally indebted to the general will, that abstract category that is usually abused for not being practical, applicable or specific.[1]

It is important, however, to note that Rousseau saw no such "hidden hand". He did not believe that trade would somehow improve diplomatic relations between states. As mentioned earlier, foreign influence should be avoided and that includes the type of economics that he prescribed. Peace requires a "deliberate act of political will" and does not happen naturally (Roosevelt 1990, 99, 108). Instead, it is easier to understand the general will in contrast to difference in privileges, distinctions or achievements, which he calls the "particular will". The goal is not to have the general will erase any type of particular will, which is the implication of Arendt's argument about the scientific outlook on the general will and its hold over the nation and the market. Were it the case that each member who joins the social contract has to renounce all instances of their private interest, the implication would be a type of sovereignty that "glorifies an unbounded absolutism of the state" (Cassirer 1988, 17). This was, however, clearly not Rousseau's intention when proposing the concept, as a check on the many examples of the particular will. He did not argue, (for the abolition of other concerns altogether as if that were even conceivable) (Dent 2005, 76).

This also applies to the relationship between science and the general will; there is little in his writing that assigns an objective status to the sciences and this extends to his conception of the general will. Arendt recognizes this, with much of her view of politics revolving around the public sphere and its relation to the plurality of men in their multiplicity. Particularly *"the human condition"* is noteworthy in how it establishes a relation between science and technology of the 20th century and political theory, notably with a wide diversity of types of engagement with natural philosophy. This is exemplary given that concepts like freedom, reason and knowledge are usually approached through a contrast of 19th-century philosophical systems of thought to 20th-century critiques. Our objective, from the outset, was to follow such an example: establishing a basic understanding of the depth and range of early modern thinking as a method means whereby to retain a commitment to freedom as a meaningful category in its relation to science and technology.

Occasionally the general will is identified with modern types of statistical majoritarianism. Most known for such an approach is probably Condorcet, who shared in Rousseau's quest for a more egalitarian society, one of the first to systematically apply mathematics to social questions. This reflects a firm belief in progress, which sets him apart from Rousseau. The same applies to any attempt today to realize the general will, at last, now that there are new types of technologically mediated forms of direct involvement and procedural representation (e.g. Azuma 2014). Any such affirmations of a scientifically mediated harmony of interests contradict the critique of progress of his early work and the skepticism of the sciences as characteristic of the entirety of his legacy. This would still imply that some type of objective status is being assigned to the general will, either through a reliance on statistics in voting, opinion polls and associated complex statistics, econometrics and so on. Any such direct involvement and engagement might more easily be interpreted as a further sign of the decadence and corruption of modern society, and by extension at risk of deepening the tragedy of life in a social state of war that is fully artificial.

Indeed, this is again the "general will" as a scientific outlook that extends to the state and the market, or rather the hold of computation over our economic well-being, public opinion and, at its extreme, the fusion of technoscience and capitalism. Science and technology, with Rousseau, are dependent on everyone and everything and this is reflected in any procedure that would effectively represent the citizens as individuals with interests, a grouping of their "particular will" (Tuck 2017, 49, 50). There are many passages that describe this as a relation of science to nature. Controversially, as described in the end of Chapter 2, this occasionally results in comparisons between "the curiosity of the man of science" with the savage who remains independent and is to be envied for he "would not turn their head to watch the working of the finest machinery or all the wonders of electricity" (Rousseau 1993b, 200). Obviously this is the typical type of compliment that might appear as a denial of a shared humanity, where not all humans are understood as fully social and cultural. This is a problem, but it is also the case that it was Rousseau who decisively reversed the relation between nature and culture, making the former a hypothesis comparable to those that scientists make all the time (see Chapter 2).

It is clearly problematic to naturalize human beings, even if it's phrased in romantic terms. Nonetheless, this remains common practice today, also in STS, albeit with sophisticated disclaimers. With Rousseau, at least, the end-result is the affirmation of the thoroughly artificial and modernist character of society, of restricted polities, whether they in terms of countries or any other political association. It is in its tragic sense that the "scientific" view of the general will needs to be affirmed. This is the incarnation that ends up celebrating the individualism of "hero-scientists", as the source of creative imagination, and which alone transcends the laws of nature, making cumbersome social norms and conventions fade away. Self-evidently this outlook will comprise education, of Emile, Julie and ours, as well any pedagogical framework, alongside the nation and the market. Each is a further indication of how the tragedy has unfolded; the scientific outlook on state, market and education have long since merged, when considering the hold of the individualism that is seen as characteristic of today's hero-scientists-business men over Rousseau's own attempts in fiction to personify modern society's "intrusion upon an innermost region in man which until then had needed no special protection" (Arendt 1997, 39).

Equality and governing

The general will is most easily understood in contrast to the "arbitrariness" that characterizes the experience of freedom among men wherever they live. This is pretty straightforward in that Rousseau seeks to confront any type of model that was still tied to types of divided sovereignty, whether in terms of a kingdom, aristocratic rulers, strongmen or otherwise (Rosenfeld 1987, 103–106). Accordingly, sovereignty has been established but based on an agreement by some, some combination of particular interests whereby authority was transferred to a person who thereby holds absolute or limited power. The problem with this model is that it makes freedom arbitrary, something that you might or might not be granted, based on favoritism or whatever those in power describe and enforce as the law.

Consequently the "general will", as a method, seeks to address how life in a modern society comes with the loss of freedom, which refers both to our intrinsic freedom, the one we are born with, and its continuous loss in modern society. This includes the sciences, which is simply another area wherein freedom might be extinguished, absorbed or transformed by committees, procedure, in contracts, by representative assemblies, governments and any other associative bodies. On the other hand, the problem is not decisively resolved, there is no practical solution to be found in Rousseau's writing that would simultaneously establish freedom as based on individual rights, as a question of personal ethics and self-awareness or as the shared ability of the citizens to rule themselves through society's institutions (Simpson 2006). As Benjamin Constant already knew, Rousseau captured the myth and spirit of modern democratic government but he did so without showing how a government could be realized in practice (Fontana 1991, 19–21).[2] While this should be acknowledged, Rousseau is always concerned about the power to reduce the role of a self-governing citizenry that values civic life and the public good above all else.

There might not be any full-proof safeguards to be found in Rousseau's thought, there are still various other ways suggested whereby to prevent how associative bodies and majorities end up abusing their power. This is again about particular interests sidelining the common interest but it implies limiting government to a "subordinate function with delegated powers" (Dent 2005, 44). This is a consistent concern, at all times he shows suspicion of any rulers or human assemblies of experts or representatives that carried the appearance of sovereignty. Its implication, however, is not that somehow delegation of power is impossible, that only the active involvement of citizens is legitimate. His perspective is a different one; while a government can never represent the general will and it can never be representative of sovereignty, "it can act as the sovereign's agent in specific areas, and Rousseau had no problem with that" (Tuck 2017, 59).

Taking the example of a voting procedure the question is not just whether everyone is included or that the vote has been cast without influence. The main condition is that the proposal should be voted on to the extent that it is in accordance with the general will. This implies that no particular interests have found their way back into the procedure; when no corporate, private or other types of personal interest have influenced the procedure, we can state that "the result is authoritative" (Tuck 2017, 49). Such an explanation is again simple as long as the general will is seen in contrast to the particular will, extended to the practice of government and administration. There are some concrete instances where he applied the general will to specific examples. This includes his proposals for constitutions, of Corsica and Poland, as well as the political and social advice to his hometown of Geneva. These type of efforts instantly show how Rousseau situated his ideas in countries where he thought life could be simple enough for the citizens to act as equals.

At the same time Rousseau recognized that it would be impractical to be involved in constant assemblies and matters of government (Rosenfeld 1987, 86, 87). Notably even a small town has to unite the legislative and executive powers, shifting attention from laws that are general scope to the particular objects that are the concern of government or representative assemblies. Even in those places characterized by close ties, needed to assemble and make decisions together, the citizenry would still have to engage with a wide array of issues. These, in this practical sense, are part of the governments duty to make the laws conform to the general will, and even to "inspire the love of it"

(Rousseau 1993a, 136–137). The function of such governments is to provide "for every need of place, climate, soil, custom, neighborhood, and all the rest of the relations peculiar to the people". And even if it were practical, the implication is not constant assembly, as "it is by no means certain that its decision would be the expression of the general will". It is sufficient that it will become clear when the general will is "flouted too openly", if it is not it means that the rulers are attempting to act justly, "that is to say, most equitable" (ibid).

There is, in other words, little that would be shocking from the perspective of a country with a division of powers that includes a well-organized court system. That is to say that the laws as written rules likely to be turned into extensions of particular interests and for this reason there are already corrective mechanisms. For example judges don't simply follow rules but are to some extend expected to apply "good faith" in their rulings, both in public and private law (see Ucaryilmaz Deibel 2022). This also extends to how Rousseau believed the difficulty increases when states become less homogeneous in terms of the status and wealth of the citizens. As a country becomes more heterogeneous and populations grow to the size of tens of millions, the ties between citizens change character in ways that limit the ability of the citizenry to conceivable engage with the various topics of importance. The practical situation and his advice remain the same. Most of all, Rousseau restricts the authority of government in the sense that administration and juridical functions should as much as possible be attached to laws that only apply to all individuals equally.

Aside of these formal conditions, citizens should have the opportunity to voice approval or disapproval as well as in the ratification of general laws. To this end there need to be periodic assemblies that can be called even if they were not formally summoned (Rosenfeld 1987, 91, 92). The continued existence of any delegated bodies depends on the citizens approval, which includes any conclusions drawn as these are legitimate only to the extent that they should have been intended to match those reached if the whole of the population would have been consulted. Aside of such advice, however, he did not seek to decisively resolve the type of problems of law and government that remain commonplace in many modern states that formally recognize that the citizens are equal, as an abstract principle and with various degrees of success when it comes its practice. Such "success" is part of Rousseau's legacy, as it was him who categorically sought to move the social contract away from individual self-interest, dependence on others and laws that benefit the few. Instead, he proposes a political organization wherein each member benefits from the common force of the association, which goes some of the way to any location today where it's more than the letter of the law in the wrong hand, where effectively citizens "respect each other" to the point that it is effective as a guarantee for each other's security in its broadest sense.

Crucially, however, these practical dimensions are secondary to how unlikely Rousseau considered such an outcome. Such proposals should always be seen in the context of how easily subsets of people grab on to power and establish that others are not to be considered as their equals. Rousseau never claims this type of power struggle will end, nor that his version of the social contract can establish how to rule legitimately once and for all. What he does argue with some force is that any other type of authority loses legitimacy whenever it pursues its particular interest as opposed to a common interest and the need to apply the law to the citizens equally. Hereby Rousseau's "general will" is rightfully the precursors of all subsequent attempts to transcend arbitrary types of freedom by asking for the submission of the individual to a strict and inviolable law, which the individual freely "erects over himself" (see Cassirer 1988, 18). This refers to a Kantian reading of Rousseau, with Kant famously acknowledging the inspiration, and Rousseau appearing as "decisive for the first germs of Kant's critical idea of reason" (Shell and Vekley 2017, 207). Its implication, however, is that the concept reaches a scale that Rousseau was extremely hesitant about; the general will is no longer tied to the type of political associations he saw as its key setting and the limited implication he considered feasible is, anachronistically, turned into an early inspiration for Kantian cosmopolitanism, as well as Hegelian affirmations of "the nation" (Neuhauser 2003).

Pacts and properties

Within the context of Rousseau's observations on government, it is not difficult to imagine Rousseau's view of the general will as a principled stance on property. Property is closely tied to particular interests and to any type of claims on power that thereby establishes that others are not equal. This holds, however, only for the loss of freedom, the tragic part; there's a notable difference when considering property as legitimate within the context of Rousseau's social contract theory.

The most famous passage on property denounces the man who enclosed a piece of land; from "how many crimes, wars, and murders" might we have been saved by "pulling up the stakes or filling up the ditch". The tragedy began because the impostor got away with it, convinced the others of his claim and made them "forget that the fruits of the earth belong to us all" (Rousseau 2003, 84). This critique, from *the discourse on inequality*, is often seen as a precursor to Marx, notably his discussions on the enclosure of the English commons, and, by extension, to our understanding of global capitalism and the field of political economy (see Marx 1990; Harvey 1996). A few sentences later, however, he refers to how things "could no longer continue as they were", how "considerable progress" and "knowledge and indus-try" were needed to make enclosure even possible. Consequently, we are once again to understand Rousseau's view tragically, it is yet another observation on the loss of natural equality and the inevitable fall into societal decadence.

The predicament is, therefore, not a principled or ideological stance on the evil of property, the glory of common property, or some type of insistence on the morality of sharing and altruism. Rousseau is explain-ing how initially there used to be no property rights; freedom in nature meant that initially at least there was no law, and rights did not exist yet, as long as there was no need. Theoretically this implies that the thesis that there are "natural rights to property" is rejected (the posi-tion held by Hugo Grotius and John Locke but not Hobbes), practically and formally property rights are characteristic of the journey out of nature and, therefore, social characteristics even when they precede any type of social contract. It is, of course, practically possible to claim objects or domains in nature – either simply out of need or as the type of accumulation that becomes possible through early encounters with others. However, these are not "rights" in any sense either because they lack a social contract, or they are instated by way of a type of divided sovereignty, a social contract wherein men are not equal.

Rousseau agrees with Locke that possession in nature is charac-terized by its insecurity. We might see the fruits of the earth as being common to all, but whatever you claim as yours can be taken away by anyone else. Famously, Locke argued that what is the result of the labor of your body is yours. This provides Locke a means of privat-izing what is common to all, by way of granting rights to the fruits of labor. This was quite direct, as it was applied to the parts of the world being colonized, which he consistently characterized as a state of nature (Tuck 1999). Accordingly, there is the desire to safeguard prop-erty and, with Locke, this is part of the terms of the social contract. The sovereign is, in this formula, legitimate because of the need and ability to provide security and this includes natural property rights. The legitimacy of rule thereby comes to depend on the ability to guar-antee property rights, and by extension the distribution of wealth, pri-vate land, wages and so on.

Rousseau, by contrast, started from a position of initial equality – a condition wherein no individual holds a natural right over another, but such equality cannot last when we begin to encounter others. Giving up our freedom was not a choice, it was imposed on us through the tragic conditions we find ourselves in, and the journey out of nature is inhospitable enough for us to join together for a wide variety of reasons. Some of us might choose to preserve ourselves, some care for the suffering of others, others require guarantees for their property and so on. It does not matter, we all have to come to the realization that the meaning of freedom depends on a social contract that establishes the necessary variety of reasons for each of us to join voluntarily. Accordingly, it is only such a "free" social contract that can authorize and legitimize property rights. It's not surprising, therefore, to find passages where Rousseau argues in favor of private property, or, rather, where he explains that property rights are part of the social contract.

For example there is a dedicated chapter on "real property", in *the Social Contract*, where he describes that each member of the community gives up whatever is in their possession when the social contract comes into effect (Rousseau 2003, 196).[3] Its implication is that private property has suddenly gone from being an integral part of the tragedy – unjust in nature and without any positive effect on the society that came out of it – to being an integral part of the solution. It is integral to his social contract theory and how it revolves around the concept of the general will. After all, everyone has to give up everything when they join the social contract. We come to realize, out of our particular interest, that we have to agree, giving up everything to the community so that nothing falls outside – not possessions, powers or natural freedoms.

Most, of course, might have been motivated to join because we require protection from the insecurity that otherwise characterizes relations to others in modern societies. A few of us, maybe, might prefer a Kantian reading of Emile, in that you rationally reach the conclusion that your freedom requires that you bind yourself without reservation to the common benefit in political society. What, however, about those who were better off than others? Why would you, in such a starting position, agree to be equal? The reason, we already discussed, is that of Locke: the legitimization of property, its inclusion into the security provided by a political association, through the establishment of a legitimate type of sovereignty.

On the one hand, the social contract is the only way to secure the possessions that were claimed before it comes into effect. Hence the rich man voluntarily joins in with the rest of us. The rich agree to join the social contract because this is the only way their possessions are ever going to be legitimized in the eyes of the others in the community. On the other hand, Rousseau insists that the rich give up their claims to the community and that under the general will their possessions are returned to them. Hereby it is established that through the general will there are limits on the rights property, unlike the seizure of lands, possessions taken by strength alone that exist without such a social contract. As mentioned in Chapter 2, Rousseau is keenly aware of the colonial outlook of Locke and Grotius, how their work legitimized such plunder. This includes detailed discussions of force and occupancy as fundamentally illegitimate when it comes to the foundations of property rights. Additionally, the type of society that he thought was suitable for his type of social contract would have no colonial ambitions, being inward-looking and staying away from more assertive societies, avoiding dependencies. This includes property, as it will not be able to operate like a private domain, shielded from the community, for example, when dealing with taxation, passing private land or other such conflicts between the particular and the general will.

Only the Rousseauian social contract can be that foundation and the overarching point is that Rousseau establishes that everyone has to have an interest, a selfish reason, to acknowledge and accept a duty to the common good. They do so out of necessity, and not out of morality (like with Kant). Those in a strong position join alongside those that are not, and this implies that both have to subject themselves to the control and direction of the others. It is, therefore, not as an ethical imperative that the social contract has to be accepted. Rousseau sought to exchange the idea of individual self-interest, dependence on others and laws that benefit the few, with a political organization wherein each member benefits from the common force of the association. The reason they join freely is because they realize they are in a tragedy and it is simply advantageous to agree to declare their loyalty to the whole of the people, or, rather, to the rule of the "general will". The individual is not asked for any type of "moral conversion", as the motivation for the acceptance of the condition of equality. In effect, man only "only obeys himself in this act of union (Cassirer 1988, 19) and "he gets back what he gives and more besides" (Althusser 1988, 105).

This is key to understanding how Rousseau never eliminates difference in privileges, distinctions or achievements, making it clear that "the particular will" is legitimate exactly on condition of being a member of the whole community. No one lost anything they had, everyone gains security and authority is established as more legitimate that any model based on a divided sovereignty can provide. This is the full formula again: that sovereignty is no longer based on the agreement of some, but that it is indivisible. This began with a formal equality criterion, the start of the formula are the various descriptions of nature that demonstrate how inequality comes to characterize the journey. Yet freedom is inalienable, even though the journey implies that natural freedom is lost constantly. The only way out, an escape from this tragedy, is a state of society ruled, formally, by all equally. Hereby the basic equality of those who joined the association is established as based on the conditions of legitimacy of the rule over anyone that is a member of the political association.

On the one hand, this formula is logical from starting point to conclusion, affirming how freedom was never given up, sold, or exchanged to another, a point he makes emphatically when arguing, in "the Social Contract" that "to renounce liberty is to renounce being a man, to surrender the rights of humanity and even its duties" (Rousseau 2003, 186). On the other hand, its logic is only possible because of how unanimity is invoked as a strictly formal criterion, and (again) not as a practical perspective. Everyone agrees, but you are included only if you agreed to be included. You are a citizen because you affirmed the social pact while others are not counted because they disagreed. Its implication should not be surprising as it applies to most constitutions and countries. Even when we extend the equality principle to the scale of the practical matters of government in modern nations inhabited by millions, the end-result is still the rule by the general is logically tied to a polity that has to remain restricted.

This was not an intellectual oversight; there's little to indicate that Rousseau sincerely sought to open up the criteria for citizenship; assemblies and representation apply to citizens only while non-citizens within the territory covered by the social pact have to live by its laws. And even if we overlook this viewpoint, there is always this point where any social contract theories impose limits on who counts as "the people", who is excluded or is encouraged not to participate, what the insiders are supposed to be like. The Marxist position would be to point out exactly this: that social contract theories are meant to "suppress the existence of social classes" (Althusser 1988, 116, 117). Also practically unanimity will not be something that can be expected as the outcome of any kind of bio-constitutional analysis, given that the meaning of freedom would have to somehow arise directly out of how code and information are getting under our skin – into our heads, our hearts, and escaping out of our hands.

However, there is another fundamental problem to consider. Surely authority should not be legitimate only in the case of a social contract that can accommodate every single person everywhere. This would not only imply that each associate of a polity would have chosen to join voluntarily but also that new members will make the same choice voluntarily and that old ones re-commit, continuously. What does it mean, therefore, to state that sovereignty is indivisible? What other conclusion can there be, within the context of Rousseau's work, than that the polity has to be kept restricted for our relation to freedom to retain its meaning?

Method and truth

Our problem with the social contract is not that Rousseauian democratic ideals are stuck in the political world of his time. Nor should we mind too much about how the world of his time did not conform to his ideals and how his democratic ideals do not conform to our contemporary world of nation-states. There is another direction to take: interpreting the concept of the general will methodologically (that "if" from before). Accordingly, we affirm the tragedy but without the need to claim empirical truth, and in the knowledge that our claim is there already. After all, we're following Rousseau's method, which means we are doing research as a meaningful approach to establishing the freedom of those that join in. For that to be possible we need to establish what might qualify as the general will in relation to the legitimacy of rule in restrictive polities. Consequently, the method is not a comparative approach to national cultures of various kinds, scientific or otherwise. The examination needs to establish a relation to Rousseau's social contracting that is at once less detached – providing specificity as a method – while establishing a more comprehensive approach to the precarious relations between society and nature.

The latter refers to Latour's hybrids as a "reversal of the Kantian solution", which is to say that the reality of nature is not about "a derived, phenomenal reality". Instead, nature and society are "derived from the primordial activities of something in the middle" (Collin 2011, 131). After having engaged with some detail with Rousseau's social contract and its many implications, it is still possible to be critical about origin stories, to investigate foundations, and include nature in our analysis of science and technology. What changes, first, is the relationship to the credibility of facts and their claims on universality, inclusivity, accountability and the various procedures involved. Subsequently the approach can turn to the responsibility over, and the democratization of science and technology.

Clearly Rousseau mistrusted the authority of the sciences and this applies directly to our own experience of living in times characterized by a questioning of the authority derived from formalized types of knowledge, learning and (applied) arts. Rousseau, like Hobbes, would agree that scientific societies often lay a claim on the truth in ways that further their own particular interests, to the point of challenging the common good. The sciences exercise their own authority over what is truth, over what knowledge is credible and how instruments and laboratories establish facts. Such a position is nothing that is disturbing, given today's approaches to the objectivity of facts, the methods involved and their social organization faces a wide variety of criticisms (see Chapter 2).

However, what we have to acknowledge (as a "wrong method") is the nostalgic yearning for a mythical time when there was still integrity. What this suggest is that there was, not so long ago, a time when the facts were still the facts while the solution to the chaos characteristic of the present is a return to order by once again strengthening the institutions of knowledge. Such a return to order is a question of authority over knowledge that echoes the formula of an omnipresent and omnicompetent sovereign. Truth is losing ground, and so the facts have to be kept in order, otherwise power and knowledge might become indistinguishable. They have to be assigned their right place and their relationship to the truth should be kept transparent. Accordingly, authority has to be established by keeping control over private, corporate and foreign interests that are ready to meddle in our privacy, elections, public opinion or otherwise destabilize the order of society. Hereby the (wrong) method turns into a linear appeal for authority, which is about keeping at bay dubious truth-claims, whether those of flat-earthers, vaxers, big foot enthusiasts as well as revelations by scientists that match the interests of big tabaco, oil, pharma, the chemical industry and so on.

Another such (wrong) method applies to the type of escapism that defines "nature" in terms of the various technological attempts to overcome disease, old age and other types of physical decay. Such types of escapism correspond to another type of call for authority, one that is exercised over the development and management of technologies that provide us with more health, wealth, well-being, strength, beauty, intelligence and so on (see Tamminen and Deibel 2019). Also, this type of order once again invokes a Hobbesian pact: your natural freedom is worthless (aren't you afraid, don't you suffer?), so why not leave it? You transfer your rights to another entity who takes over the responsibility for your safety, your security and will inform you of your relation to the truth.

People consent to many things, and so do we, especially when it comes to our own relation to science and technology, to facts and their complex relation to the devices we surround ourselves with and depend on. Its implication, however, is not any type of deal or pact that is legitimate in the way that the Rousseauian social contract aims to be. Rather, such agreements reflect the "wrong methods", which we get to keep in our analysis, as a demonstration of the deepening of the tragedy. In this case, our predicament is our technologically mediated experience of escaping our own nature as well as reinforcing the type of social order that revolves around controlling what we know, how we learn and how we gain our skills. Accordingly, there are many hybrids all around us, more and more of them surround us, and we acknowledge how we invariably become nature-culture hybrids ourselves. Yet, there's no need to see such hybrid combinations of facts and social order as if there only ever was a straightforward belief in the indisputable realism of scientific facts with trustworthy academic institutions as their place of origin.

Does the authority of truth depend on the belief that it existed in the past, or, at least, that it was once taken at face value? Its familiar shape would revolve around overcoming a separation of facts from society, unleashing the potential for democratization of science and technology, as a key site for democratic agency. This is the implication of the STS literature on the role of public participation in scientific and technological decisions-making and design. If we don't, the next generation will inevitably grow up and consider such types of "dependencies on men" as a given, how could it be otherwise? Yet, this boundary between science and society is not characteristic of Rousseau's tragic conception of freedom. Indeed, such a separation is part of the corruption of society, integral to how it functions in the present, our present wherein crisis, catastrophe and dystopic conclusions are a constant. This remains the case when we, idealistically, look only forward and associate ourselves, in Rousseau's shadow, with a more direct democracy, invoking idealized polities wherein everyone is equally a citizen and extending this to our concerns over science in society.

Rousseau's shadow includes our own pedagogical framework, the one that the next generation might grow into. A different type of boundary work needs to be affirmed, as an indissoluble part of modernity (and of STS courses). We know understand that Rousseau's method does not revolve around claims on empirical reality and truth, either in (critical) realist or constructivist terms. The latter maintains its place in our acknowledgment of there being multiple and precarious relations to nature, while the former finds purpose in the need to "judge properly our own present state" (Rousseau 2003, 44). Therefore, the point is not that such "judgment" implies a near continuous commitment to decisive ontological claims. Nor is there a performative imperative, of practice and being practical (from mainstream STS). Such stances make little sense within the context of all the dimensions of Rousseau's work. When talking about Emile he referred to the stages of development of childhood, to the law of nations, the sciences and more. Invariably such "supplements" (see Derrida 1988), "discrepancies" (Althusser 1988), "inferences" (Collins 2004) or "hybridizations" (Latour 1993) are a necessary means to an end; open to be challenged, as realism, as empirical work and yet decisive is only whether and how their relation to the overall ontological claim remains intact.

Our own attempt at Enlightenment

While Rousseau's method departs from a commitment to a relativistic "if", its aim is a meaningful relation to freedom. To that end it engages closely with the "wrong methods" of his predecessors. Accordingly there are various types of naive naturalism as well as more nuanced ones. Methodologically there's, therefore, no reason to conclude that there is nothing else than to re-enact the same types of laws of nature, the ones that seek to retain their hold over the histories and languages of freedom as established ways of thinking and acting politically.

After all, the precarious relations of nature to the sciences are once again anti-deterministic and anti-essentialistic in their origin (see Chapter 2). The various origin myths wherein we might encounter a state of estranged independence, moralistic selflessness or a fundamental loneliness are hypotheses and powerful only as fictional affirmations of open futures wherein there is, somehow, only space for narratives that promise reflective combinations of (empirical) science and (empirical) society. And yet, there is little freedom in such a scenario, with all of us as witnesses to democracy on the decline, the kind wherein freedom is optional or superfluous relative to how code and information are getting under our skin – into our heads, our hearts and escaping out of our hands.

Let's say, therefore, that we go out of our way to code something truly enlightened that actually slows down the flood of new things. Something like the freedom-machine of Chapter 1, which is able to let us keep some of our freedoms – at least those that are compatible to the benchmarks of 21st-century software engineering. Maybe you're one of these people, one of the good ones, the ones whose designs are different. Good for you, you take a lot of time addressing all the ethical concerns that were raised as faithfully as you can. Congratulations, you went through the trouble of staying on the moral side of the line, even acknowledging it has many dimensions and contexts. Yet, we learned a lot about our desire to be free: aren't we still simply running towards our chains? Isn't it still the case that we are imagining some type of historical determinism that we will have to confront, or maybe it is another counter-Enlightenment to defend, or it can even be one wherein God has been replaced with a certain type of Science-Nature? Have your pick, invoke some type of journey from a known beginning (paradise/nature) to a known ending (the fall/a tyranny of science in society). Whatever it is, it travels with you and thereby reminds us of how no one is more dependent on everything and everyone than scientists are, as Rousseau asserts.

Clearly we need to find our distance from how certain "people" do their things (as dependencies on people). Rousseau's shadow is long, particularly when the champions of direct democracy show up, and they are never far away when considering the numerous policy processes that seek to mobilize the direct involvement of a diversity of stakeholders and engaged participants in scientific projects and into a continuous interdisciplinary dialogue among industry, science and civil society (see Nahuis et al. 2007). Such projects invoke an unspecified demos in how they aim to derive strength from the ability to represent and accommodate converging interests, diverse points of view and multiple understandings of the transition to more responsible, sustainable or ethical standards. Naturally it turns out to be a problem that criteria like inclusiveness, participation and engagement are regularly seen as obligations rather than preconditions for responsible research (Nahuis and van Lente 2008; Wynn 2008; Stilgroe et al. 2013). Studies of the public understanding of science and related approaches typically conclude that citizens involvement with science and technology is affirmed is desirable but without fundamentally challenging the role of experts, their place in society and the authority derived from establishing facts or organizing the democratization of knowledge.

Accordingly there are many attempts at bridging, patching and joining together what is assumed as separated. And usually, the symbioses of things and people is presented as a democratic imperative. Projects and programs incorporate ethics or social dimensions and citizens should be encouraged to freely choose engagement with the sciences but on the premise that the facts of the matter will not be fully included in any type of decision making over the matters that affect all of us, include all of us and lead us toward a shared future. However, inclusion is not truly possible on these grounds; not on the assumption that progress has a positive effect on public morality, which was what Rousseau challenged first. Wherever consensus-seeking with stakeholders and the public is being performed, Rousseau's critique of progress applies; the one from "the first discourse" as well as to whatever particular interest uses the power of government to introduce benevolent measures, as gifts or other types of support that subsequently turn into a repressive mechanism (Dent 2005, 94).

Therefore, the focus should remain on the legitimacy of such interventions, on what type of social contract this invokes. Certainly, we need public involvement in science and technology and strengthen the democratic legitimacy of science. Perhaps this requires expertise, and, occasionally, this might imply that there are limits, that proper boundaries need to be drawn so that the citizen-scientist does not overwhelm the expert, as an old STS debate suggested. Yet, this would not be enough, as such an adjustment would still keep us grasping at anything that might optimistically invoke a restoration of histories, networks, forums, agoras, parliaments and instruments that are needed "to compose a common world progressively" (Latour 1993). This version of the "strange" multiplicity, as characteristic of today's social contract, does little in terms of giving direction, even it claims to be putting us back on our feet, letting us re-imagine new beginnings, become an unforeseen movement or have our actions take on the intended meaning (see Tully 1995). The situation might be bio-constitutional, mixing strange laws with strange lives, and yet this simply confirms how the particular will rules, unchallenged. We are left without an objective, without any type of ontological claim to guide our method, one wherein we investigate, practically, the different type of pact that this implies.

Freedom machines and beyond

Isn't it something: finding yourself appealing to the potential of free software to reject how commercialized software and hardware energize the constant flood of digitally altered behavioral patterns, ever more integrated with the knowledge about our genes and brain functions? Oftentimes such devices are tied to hopefully marketed platforms for curing diseases, ending hunger, unnecessary suffering or the transition to green and clean energy. Who is not sympathetic when such naive naturalism revolves around children with catastrophic illnesses, would-be mothers, the inhabitants of some type of ecological collapse, or the animals that don't need to suffer if only we eat synthetic meat? And even if it's not that altruistic: who does not at least occasionally choose their own chains, joining in with the rest of the naive naturalists who similarly desire to be affluent, living in sustainable societies and in perfect harmony with each other and nature?

We might, by contrast, identify freedom with software, and how it found its full expression in the circles that were created around self-sufficient and independent-minded types like Stallman and Thorvalds. The claim that there should be "free" software soon had its many practical examples; in a short amount of time it gave rise to, morphed into and was absorbed into something much larger and different. Quickly a way of working came to co-exist with the ethical claim. It is as if we are still in Rousseau's state of nature, with the origin of the freedom-machine referring to the period of isolation, when there was an equality condition being establish. While it did not last, given the inequality that characterized the world of intellectual property in its relation to software, it still delivered the freedom-machine. The result was a functionally superior system, superior to the ones protected by countless intellectual properties. It is by comparison "free", a system kept up to date in terms of its functionality by a user community and much simpler in how it runs basic services.

With it came the type of security and simplicity we can expect in relative isolation. After all, its programming is more transparent, anyone can make contributions and check each other's work. This keeps contributors honest and this honesty is integral to its greater security, less bugs, easier to fix and, thereby, more difficult to hack. Unsurprisingly, the freedom-machine won, these features made it the system of choice in many areas, running in the background of the world of business, of governments, your entertainment and whatever you do that is not free of charge. This is the new deal, after it turned out that selling source code is hard, even with the global push to expand intellectual property rules, to strengthen their relation to the rest of us. To legitimize property – what source code is theirs and not yours, who decides how to use it – requires that we recognize how much easier it is to distribute it, copy-pasting it with a few clicks. Consequently, it is not given that a program can be packaged in a way that results in payments; whenever software is involved, this requires persuasion or, even better, some type of legitimacy for the business model.

We all know this, it's the way it is for whomever has stumbled across unauthorized files of music, films, games, programs or even entire Windows and Mac operating systems. Downloading these out of convenience can testify to this, even if it is just an exception. Convenience and persuasion, however, are not freedom – no matter how effective the machine-learning has become that drives the smart devices that will deliver us from our boredom, distracting us from the tedium of our own tragedies. Legitimacy of property, with Rousseau, requires a social contract that no longer reflects some type of particular will, the interest of some sovereign and how rule is established over others. This part is simple. At the very least it is an example of the particular will when someone, coders included, consider themselves as "free" while living in isolation or surrounded by each other in their preferred technological habitats, gated communities, corporate techno-parks or university campuses.

These settings show that we are leaving behind where it started – the self-sufficiency and independence of the early days. Such a dependency on things had its day and it could even be argued that as we grow up, we are seeking to maintain its spirit as long as possible. But we do so only for the sake of Emile's education, or for the sake of our own little cute cyborgs, growing up to live and flourish in our own modern societies. We know it, it will not last; as a relation to freedom, it can no longer retain the same meaning as it used to. Some might tenaciously continue to imagine themselves as coding their way to freedom, in control of their own destinies from behind a computer screen. This is to be expected and might even be actively encouraged, reflecting how a social pact has been struck: one of those that preceded Rousseau where only some conditions are imposed on you while a few of us get to keep their natural rights because they are most special than the rest of us. Why don't you keep developing those freedom-machines, as long as you don't ask any question about what type of society it will lead us to? You'll clearly still be free once it has driven us straight into the fourth industrial revolution, into the internet of things, the posthuman, the singularity, the Metaverse or any of the other types of speculative future.

Each of these marvels will at some point, once again, be presented to us in terms of freedom, decentralizing production, democratizing how knowledge is generated algorithmically – exclusively and aimed at our belief in our inherent altruism. The march goes on, with freedom-machines as stand-ins for our deliverance, helping us reach the truth based on mystical algorithms that are running smart devices and our synthetic DNA. What a relief: human nature turned out to be uncorrupted after all and can be relied upon as a natural foundation when imagining a new society – this time fully compliant with a narrow liberal sense that we should be free individuals. After all, it can now be guaranteed by new types of machines that we can have faith in. What is the alternative? Is there anything else than to retreat into other type of naive naturalism: the one that will, at last, allow us to finally draw that ethical line in the sand? This time the line will hold, because there's nothing around that is powerful enough to overtake an agreement that would stop the various types of interventions into our self-awareness as autonomous subjects, equipped with rights, reason and dignity (Fukuyama 2002, see also Habermas 2003).

Who knew? Rousseau should have been read as a romantic or a counter-revolutionary after all – if we go along with the idea that the nostalgic longing for a return to a "real" human nature was right all along. Our human-centered ethics prevailed despite of it all; and it even took the shape of a full renewal of our age-old believe in universal knowledge as the foundation of liberal order. Only a small change was needed, which is that freedom, as our given right, required special machinery to operate as the foundation of our societies and ensure that everything remains as promised. Passing from Dante's vision of a cosmological order wherein everyone is assigned the right place, we arrive at Diderot's Encyclopédie – when the facts were still to be presented as the facts – to the universal knowledge of culture and life of Wikipedia, and the digital instructions of each biological feedback loop that will run on a new incarnation of superior code, compiled and generated by millions of collaborating volunteers.

But we are not to be overly-utopian; after all Rousseau's conception of freedom should be tragic. For this reason, there was only a tentative affirmation of the ways wherein we might look forward to a time wherein a new type of freedom could be realized. Despite of the difficult circumstances we find ourselves in, Rousseau was no breaker of chains – unlike the revolutionaries that he inspired soon after, the workers with nothing to lose that followed and so on. Such were the tragedies of centuries past; it's ours today and we are similarly invested our own predicament, pondering our own escape, as escape we must. Inevitably a new generation is always gaining whatever awareness available of its own agency, within the confines of societies that are not equal. There are differences that matter. They might choose to challenge the legitimacy of the social pact as it exists, aiming to establish a condition of equality. Or they might choose to recreate the restrictions of their polity for their own benefit. Sometimes it might even let the rest of us retains some notion of freedoms, and perhaps there are some components where a common interest, a general will even, can be detected.

Emile did not grow up to escape the inequalities of society, this was never the goal; rather than "to relegate him to the depths of the woods, he was to be prevented from letting himself get carried away by either the passions or the opinions of men" (Rousseau 1993, 260). Like him, every educational model comes with a hierarchy between student and teacher, an inequality introduced into education and it comes with considerable authority and at least a sliver of conformism projected onto the child's behavior. Such hierarchies cannot be erased simply by opting for a more progressive education that seeks to provide settings wherein the child's self-awareness does not rely on differences in wealth, background or any other example of inequality. The problem has not changed, it is still about helping new generations escape from these inequalities and this requires experimentation, finding new formulas whereby a realization is gained about the meaning of inequality. This requires a commitment to idealized societies or polities, critically and reflexively, but as a necessary part of the educational task of preserving the future possibility of action as it is being overwhelmed by the demands of a world that is in constant need of renewal (see Lelja 2018).

Each generation, like Emile, ends up in the situation where it has to make its decision about whether the social pact was joined freely, perhaps imposing demands that reflect the necessities of their predicament. Its implication is not that we simply aim for maturity, joining our communities with some type of constructive attitude, a sense of nuance, an appreciation of its complexity and so on. This is insufficient, a Rousseauian method is aimed at becoming one of those "who are to be", by making sure that "empirical reality does not obscure the ideal possibilities which are to be established" (Cassirer 1989, 122, 123). Despite of the tragic scenario, which makes the task difficult and much less practical, it's a simple formula. Emile is to become the independent minded individual who would not be persuaded by luxury, comfort or any other private interests. Only such citizens that rule themselves can keep a close enough eye on the authorities and even on other countries, directing their strength to a shared sense of security and well-being in a world of predatory states, and the particular will that usually governs such territorial sovereignties.

As soon as we assert a return to the Rousseauian demos for a new generation, the claims on freedom become much less abstract; they are about your relations to others, your offspring, the substance of the new community you want for the people you care about, the presence and strength of the common good and mutual aid among citizens. When our self-interest in freedom is re-asserted, practically, there is instantly a balance to be struck, a pact to affirm, with conformity and control as unavoidable features. Our education, like that of Emile, took place in a world that was always riddled with the old contradictions of divided sovereignties, including the authority of our own tutors, who thereby got to personify the rule of the monologic of sovereignty and the imposition of a homogeneity of order. This matters greatly, practically, but is not decisive when presented with the choice, whether or not to affirm that we should organize the polity around a commitment to being free.

Accordingly Rousseau spells out our experience of our modernity, of our transcendence of nature. We are rushing once more towards our chains and, with Rousseau, the only alternative that ends the tragedy is if we are able to "unite and choose our chains". The model for the transcendence of nature is one that revolves around polities where an ideal of freedom merges with our experiences, our self-interested claims derived from some of the many precarious relations to nature that characterize our tragedy. While this means there is foundationalism – many types at once and naive naturalism/s everywhere – the path is neither linear nor is there any single authority figure or framework to tell us how to take responsibility over our world. This includes the democratization of science and technology to the extent that it reflects the scientific outlook on the general will, the one that Arendt discussed so elegantly.

We know what to expect: policies or projects that establish social harmony are integral to mass societies that require a constitutional ordering of nature, as a managerial matter, and seek to continually absorb its social groups and classes into the whole. Accordingly the young will get surrounded by educators, as our shared interest, but also with some type of authority to determine their relationship to any type of hope for a shared future of change, openness and horizontality. At best the pedagogical framework we can imagine ends up preparing us, like Emile, to join a society that is highly unequal, with mostly only a halfheartedly established vision of the society that might come to exist. Yet, this does not diminish its implication: that it is worth seeking to retain a "dependency on things" as long as possible but only on the understanding that this is hypothetical or even fictional in its relation to the necessity of things, and with an ontological claim guiding our method.

Freedom and security

Let's return, in closing, to the proximity of Rousseau to Hobbes in terms of the relation between freedom and security. Why not take, as its illustration, "Fri-post: the free email association"?

This is a very small community in Sweden of free software enthusiasts that have setup email communication infrastructure with as its objective to give users control over their data, which is described on their site as "user freedom". This infrastructure operates as a distributed network of a variety of servers as well as the maintenance of reliable email services involving mail user and transfer agents. All of these technical elements are safeguarded further by the democratic participation in the community of volunteers that is committed to the transparency of the procedures. Seen through a Rousseauian lens we might choose to see this in terms of an engaged citizenry, involving individuals and having them take responsibility together for their own privacy shows a merger of types of freedom – to associate, of speech as well as its relation to security. The latter is straightforward enough in how our ability to communicate has to be build and safeguarded technically and democratically through an engaged community.

Accordingly Fri-post appears as an extension of the simple deal, again the same *"U-R-Free"* that the "freedom-machine" of chapter one was using. Fri-post simply extends how free software licenses impose on its users that the code has to remain free in legal terms and users are often persuaded to join by the extensive online support. Surely this tiny illustration cannot in any conventional sense qualify as a "case", one that would allow for the tragic relationship of software and freedom to be reconsidered comprehensively, or to demonstrate that another reality is possible than the one discussed in terms of "freedom-machines". The same applies when adding other such projects, of which there are many. Each of which could be shown to maintain highly detailed and worked-through forums where users contribute by making their concerns known, give feedback and filing errors. Such posts or help ends up shaping the users into contributors by providing exactly the type of right descriptions to beginners, generalists and niche specialists alike. Ideally such a learning experience, as a pedagogical framework, turns the abstract value of being free – of openness to other things and mutual respect of different relations to the object– into a practical skill.

To refer to a specific illustration, Fri-post for example, can therefore be simply functional; each illustration is included to the extent that they are theoretically useful, helpful in the development of an application of Rousseau's method step-by-step. They are not-a-case study (see Chapter 2), as they are clearly lacking when it comes to empiricism, deliberately so. They could be discussed in empirical terms, but within the context of this methodology they should be deliberately lacking in empiricism. Otherwise the intricacies of the case ends up as a distraction from the objective, a sufficient and meaningful understanding of freedom; it would get in the way of a methodological relationship to the goal of making tangible how "nature-cultures" of this kind remain dependent on the kind of "things" that might be needed to set us free.

This was useful already, this single illustration focused our attention on initiatives born out of necessity – out of a dependency on things that we can methodologically determine as worth investing ourselves in, as it revolves around establishing ways of doing things together and cultivating an appreciation of freedom as something that the digital world requires.

This illustration (and not-a-case study) is also immediately useful when observing the relation and relative differences in comparison to the freedom machine. The latter refers to freedom in terms of source code that is made available freely, with a community of freely associating individuals that is organized around programs and operating systems that are functional without any of the restrictions as the result of the normalization of intellectual property. Fri-post might fit within this definition, but its members attempt to sustain themselves as a local community where the free association comprises very different types of skills and skill-levels. It's not special in this regard, as there are many community networks controlled by members in contrast to the opacity of today's centralized and networked platforms. Perhaps, Fri-post is more useful in that it is fully functional while many others are temporary DIY or artistic expression of dissent or provocation. Nonetheless, it would not contribute to review such initiatives comprehensively (see Dragona and Charitos 2016) as it is sufficient that Fri-post has sought to move away from the strictly technical criteria of user-centered approach to privacy or stability and towards security as something anchored in a democratically organized community.

Hereby it shows a relationship of freedom to security, which is also a core concern of Rousseau and political theory generally. The difference matters. On the one hand, the illustration clearly foregrounds one of the key dimensions of how languages of freedom, like the one that started with the Free Software Foundation, are not invoked in isolation (as ontologically distinct) but in response to how technology is traversed and constituted by relations of power. This can be established without to many complications, but it leaves the goal of the association, freedom, as a trivial part of the analysis, taking us to the conclusion that freedom is contingent on lots of other things, but not any further (see Chapter 1). On the other hand, with a Rousseauian method we can demonstrate how the association, as a restricted polity, is renegotiating the theoretical relationship of freedom to security, and approach it empirically to the extent that it does.

This is only a single aspect, identified with a single illustration. However, it is not difficult to do the same for the relationship of freedom and truth. After all, much is at stake when it comes to our ability to imagine a future wherein there is still that same focus on freedom, a free software mentality, to misinformation and government propaganda. There are such examples, at least in intent. There should be "an intelligence agency for the people" that approaches the relation of freedom to truth at a geopolitical level with a "firewall of facts" (Higgins 2021). Perhaps such initiatives are developing, empirically speaking, but even if they are overestimated, they remain relevant as examples of experimentation. Rather than simply resorting to a flat naturalistic ontology, the very focus that is needed for such experimentation implies a commitment to the theoretical relationship between freedom and truth.

Perhaps this is an orientation that simply ends up deepening the same predicament, expanding on the freedom-machine through another incarnation of the underlying belief in progress as our savior. This is possible; it would imply that everything remains the same, even if the criteria for the application of the method have changed somewhat, further extending the functionality of the "freedom machine" and the tragic turn it has taken. What if, instead, we were to consider Fri-post and other such restricted polities as the backdrop to revise the entire deal; part of an attempt to move towards a different type of pact? What about a different type of *"U-R-Free"*, as it was called in Chapter 1? What *if* we move on from its "features" – like truth, security and property – to the more comprehensive view of the precarious relations between nature and society that characterize our method?

Each of these relations comes with claims of authority over the necessities we want to keep and those that impose themselves. These are to be taken (more) seriously, even if such realist claims are secondary within the context of our method, and by extension as a type of practical STS research. Given, from the start, is that this implies that countless avenues (re-)appear for new epistemological viewpoints from where to (re-)contextualize and (re-)materialize our commitment and duty to a shared world. The aspiration, what we need, is that thereby we begin the work of re-imagining our lives together as born out of necessity, recognizing how we have little choice but to surround the social pacts we have with our own, ones that are out there, that potentially might give them a new shape and direction.

Freedom and the natural world

The next illustration applies even more directly to our precarious relationships to the natural world. Consider again the practice of open licensing, as a basic condition of equality, but this time it is about environmentalism and agriculture.

Rousseau's equality condition does not imply that property is impossible, but it does come with limits. In its legal shape this is echoed in open licensing by the imposition of its restriction: that being allowed to use what is shared with you does not allow you to turn it into something that restricts others who share. Most typically this refers to the ability to turn part of the shared resource into something that is exclusively owned by way of copyrights. Clearly this shows that sharing still comes with authority relations, and in this regard there's always some controversy surrounding the various examples that work in this way. For example, it has been controversial how individuals can have considerable influence over projects, whether Stallman, Torvalds or otherwise. This, however, is of limited significance, as conflict simply diverts efforts away from those projects. Even if this type of "particular will" is exercised forcefully, it does so in parallel to other projects, each with their own imperfect approximation of what the general will should look like.

Such restricted polities are limited in scale but, when seen together, they have had an impact already when it comes to the functionality and development of the freedom-machine that we started with. However, our concern is with alternatives to this trajectory and in this respect our methodology is guided by the core elements of Rousseau's formula. This is more specific than simply including more examples that might illustrate how far-reaching "the dependence on things" is and how it comes in many forms. Instead, the goal would be to prioritize a variety of types of restricted polities that seeks to re-vitalize freedom. This method revolves around the identification of those examples that affirm Rousseau's commitment to exclusive polities and whose aspiration to a general will can be presented in contrast to a monologic of sovereignty and a homogeneity of order that matches some type of particular will.

The next illustration, in this sense, is another straightforward attempt to establish credibility, which is of vital significance for any of the highly specialized communities that seeks to counter the drive to privatize shared resources. After years of discussions, there are now a dozen organizations around the world each with their own free licensing in support of farmer- and community-based sharing of seeds. Its aim is an open source seed movement and attempts to "free the seed", which is clearly motivated by tragedy as it seeking to slow down the fast rate of extinction of seeds used in agriculture.[4]

Crucially there is little to indicate any direct analogy between seeds and software, beyond the usage of a version of an open license. The notion that seeds are comparable to code cause offense with farmers

and breeders; the diversity of seeds, their multiplication and livelihoods attached to them should be the kept at the center, with only
secondarily branching out into any type of coding in its relation to
modern knowledge of plant genes and genetic resources. Indeed,
one could argue that such projects can be understood as a deliberate
attempt to distance themselves from conceptions of nature as somehow becoming disembodied or decontextualized or as instantly trans-
missible across the globe as a digital technology.

While the methodology of this book is not on the minds of its participants in any practical sense, such projects still suggest the potential
for alternatives to how the language of "freedom", "openness", and
"access" is being reshaped. In this case it is concern over the commodification of crops. While this practically refers to seeds as they are bred
and planted, by extension the same applies to how crop materials have
come to be identified with a wide array of exclusive rights over many
different kinds of resources and types of knowledge – on genetic traits,
sequences, databases, source code and so forth (see Deibel 2013). Even
as markets and technologies are transforming the usage and exchange
of plant genetic materials, expressing its DNA in digital or electronic
forms, an old relationship between nature and freedom is taking a different direction.

There's a clear aspiration in these projects that goes beyond their
local circumstances. The notion of "free seed" takes on its meaning
in contrast to the various mechanisms that have some presence in
global treaties that seek to move beyond the principle of exclusion –
rather than sharing – as their constitutive basis. This includes all sorts
of "access" to technology and knowledge; these types of procedures
organize exceptions that leave the seeds as commodified as it was
before. By contrast, this movement is defined by its aim to make available seeds and plant materials to those who will reciprocally share,
while those who will not are excluded (see Deibel and Kloppenburg
2015). Clearly such projects are still marginal by comparison to the
type of agriculture wherein it is normal to do the exact opposite: to
eliminate sharing and include only those that are willing to work on
the terms and conditions of ever more intensive types of agriculture,
as integrated with (petro-)chemical industries and different types of
"assetization" (Birch and Muniesa 2020). Nonetheless, a much greater
scale can easily be imagined. It is an attempt at forging a practical
relation to freedom that is intended to persuade the millions or even
billions of farmers and the majority of breeders that are trapped in a
narrowing corporate seed market to return to community-based sharing of seeds.

This is by no means easy as farmers and breeders have become deeply embedded in industrialized agriculture. However, it is exactly for this reason that seeds should become free, rather than only accessible or technologically speaking "open" to innovation. This might take a while but it is clearly a reference to popular sovereignty in the area of food and agriculture, and by extension to a social pact. On this basis the pact might grow beyond sharing seeds in contexts that are small and marginal, but organized around restricted polities, to eventually comprise a much wider range of technologies, resources and users.

On the one hand, these norms and practices can be characterized easily in terms of Rousseau's tragic view of freedom. The movement is very much organized in response to an unfolding tragedy, a loss of freedom tied to extinction of species of crops and livelihoods with potentially large implication for society when seen in terms of what the ecological collapse of crops would imply for everyone's food security. On the other hand, the formula simply points to a freedom of association that matches how agriculture revolves around a dependency on things. Accordingly, there is a commitment to self-sufficiency and independence (as farmers and breeders), but it is also obvious that there are "dependencies" that crosscut different types of nature and societal relationships.

Remember, keeping the "dependency on things" as long as possible is a crucial part of how we are to imagine transforming out dependency on people. The relationship of nature to freedom, in this instance, is therefore not based on any type of attempt to "relegate" farmers to "depths" of their fields, naturalizing them empirically, as less dynamic, less modern and less cultural than others (see Rousseau 1993, 260). The change that is proposed implies a social pact, one that acknowledges and responds to tragic circumstances that require change, dynamically seeking to involve farmers and breeders voluntarily and freely letting users choose to join with their own crop species. Farmers and breeders, therefore, mostly join out of self-interest, even though there are rules and there is a type of behavior imposed. After all, the goal is to impose a new standard, attempting to have ever more farmers, breeders and other users (of crops as food) internalize the new norm of sharing that replaces the one that was displaced by the commodification of seeds.

Clearly there is a long road ahead, turning these types of licenses into a source of legitimacy that extends to the various other types of pacts, deals and contracts that are the norm today. To gain authority, in this sense, requires more than projects that are limited to their respective users and act as communities in specific issue areas. However, the idea to apply the norm of sharing to our relation to nature, without nostalgia, could easily be applied in many other domains. There are many related topics where the situation is comparable. Its basics apply wherever there are a multitude of individuals with creative capacities or variously sized firms who find themselves in conflict with the handful of companies that have attained a dominant market position. In Rousseauian terms this is simply the contrast between the general will and the particular will, with their respective claims on law and legitimacy. The question that it begs, however, is "how" can we consider the variety of such projects as a strategic relationship to the inherent shortcomings or existing social pacts, the ones that are once bioconstitutional in character and based on inequality.

Imagined as a precarious relationship to nature, we are still seeking to escape the tragic circumstances that characterize the increasingly intimate relation between freedom and code that we started with. Such projects are technical in nature, having long since conformed to, or lost any sense of going beyond legalistic terms and conditions of the open licenses involved. Yet, Rousseau's formula does provide something that is different: this time we are not minimally and restrictively re-imposing the social pact, following a realist understanding of Rousseau's popular sovereignty. Sometimes we make our claim on freedom, seeking to persuade those interested in the niche we are most comfortable with how legitimacy in the world of scientific and technological developments requires a strong sense of the common good. This implies, however, that we will have to create our own shared spaces, shielding some of the common world as a practical necessity.

We should do so as an affirmation of our commitment to the whole of the community, even if our restricted polity does not fit within territorial confines of social pacts or will not quite synch with the sophisticated devices we surround ourselves with. These objects were supposed to help us interact with each other, solving all our practical problems and liberating ourselves from them. Not many of us fully reject or are even able to reject this type of promise, this version of being modern, but we can at least recognize its price. That it costs freedom, even if most of the elevated opinions that today's champions of the civilized life regularly take the shape of the speculative futures that surround our devices, locking us into an ever firmer embrace, persuading us to opt-in to the hope that somehow freedom will come out of the desirable new features constantly marketed to us as gifts of progress.

Conclusion

> but all this forms a new subject that is far too vast for my narrow
> scope. I ought throughout to have kept to my limited sphere.
>
> (The Social Contract, Rousseau 2003, 309)

The final citation of this book is also the last sentence of "the social contract". It captures a situation that is applicable to its text, considering how, after three chapters, we have moved on from our starting point. We are no longer speaking only about Rousseau's tragic conception of freedom and moved to his method, outlining the relationship to doing our own research. The latter is, clearly, something that requires a much deeper commitment to individual subjects, beyond the illustrations discussed earlier. This is "too vast" for the narrow scope of this book. Doing such research has to take the shape of many types of examinations, which together would seek to reach a qualified type affirmation of idealism about a new citizenry and social contracts.

To stop here in part possible because this has been done elsewhere (see Tamminen and Deibel 2019), with extensive studies organized around the relation of political theory and STS research. It also included Rousseau and agriculture (ch. 4), as one of its topics alongside others, each part of a comparative perspective that is organized on social and political theoretical terms. This book has foregrounded in some detail that there are more reasons to recognize why Rousseauian political theory should be seen in a direct relation to STS, starting in Chapter 1, when its first sentence started by stating that "man, like software, is free". However, a comparative perspective does not have to start there. It could start with Foucault, as the earlier book did, or any other type of critique that allow for a type of STS that seeks to seriously engage with political theory, refining its empirical claims when it comes to constitutions, responsibility, democratization and so on.

By now we know this already. We should not have to keep repeating Rousseau's dictum that "man is born free" even as we know it will not last. Once more we are witnessing a tragedy in the making, the news is always about the crisis of the present that is about to overwhelm us, unless we act right now under terms and conditions of dubious character. And Rousseau does not offer an escape, he is not the utopian thinker that he is regularly made out to be nor does he let us return to nature in any romantic sense. Reading him shows us that our freedom does not arrive in a neat description of what is given to us, as our nature, available whenever we need it, as involved citizens and guaranteed by states that respect our rights. Wouldn't that be great? – that whenever we desire it, some type of technological renewal will invite us in and that will help us escape the crisis of the present and guarantee our freedom? Unless we realize our predicament, our desire to be free will remain a tragedy – one wherein we run towards a futuristic order based on the naturalistic belief in a freedom that comes in the form of the perfectly compiled code we admire so much. We get to keep our chains as the work of generosity, something that is somehow gifted in perpetuity to humanity for free.

This book, immodestly, showed that we might as well return to Rousseau, his method and his commitment to freedom, as a guide to (de-) hybridize our constitutional claims, our relation to truth, democracy, security, education and, of course, property. Also this implies a process wherein the content of nature is "subtly – or not so subtly – transformed", but, this time, not as a type of naive naturalism (see Collin 2011, 205). What we encounter in rereading Rousseau's work is not necessarily the type of overarching "Critique" that STS sought to get away from; sure, tragedy is everywhere, also when finding a methodological approach of our own naive naturalism/s. Still, there is nothing to suggest that we have to ignore STS's favorite things, like the power that exists in the orderly scientific narratives that put the structure into infrastructures, materializing time scales and memory practices across life and information (Bowker 2005). We'll still have to appreciate the boring part of politics, the consensus-making, the participatory work, as there's even a place for direct involvement, even if that's not the core of Rousseau's view of popular sovereignty.

There's no "hatred of the demos" wherever there is a critique of the ones who are doing the "tinkering, bargaining, compromise-making and coalition-forming" (Collin 2011, 144). Nor is there an idealization of the "cool scientific calculation by a group of experts" (ibid). To both cases applies Rousseau's rather conventional philosophy of science, which is simply that knowledge should not "raise itself above life" (see Cassirer 1988, 20). It is conventional, of course, from a 21st-century perspective wherein it can be pointed out that he fits easily within a space carved out by the philosophy of science of Merton and Kuhn. The same applies to the strong program of the Edinburgh school, as well as Latour's controversy with them and his subsequent popularity in fields that are barely conscious of his roots in science studies.

With Rousseau we raise the bar and we do so out of necessity. It is hard to escape our own tragedies, and the situation does not improve if we organize our relation to reality based on an imperative to commit how it is constructed, mediated, interpreted and so on. These are each interesting hypothesis, now go and make your claim within the context of a comparative perspective, inside single articles, inside books, as a collective effort. Don't forget, establish some relationship that makes transparent ontological claim that is there at the start, in the middle and the end, whether as something to escape from or to affirm. Disagree, if you like, just don't hide behind having revealed the complexity and nuance of the situation that other people should recognize in their politics, as if that is simply an affirmation of epistemology, a position that is about reflexivity and somehow not an overarching explanatory framework that is ontological in its claims.

Otherwise, there's no way forward. Continuing on would keep Hobbes in his place, as exemplary for "The Problem" – the separation of science and society, as the Leviathan that rules of over the other dualism, and that we are subjects to. At the same time, we continue insisting that these dualism don't rule us, that they never happened and maintaining that they only obscure the intricacies of power and knowledge. Yet, since the split did not happen, empirically as a matter of a real history, it must be a working hypothesis, a fiction even, which is exactly why Rousseau should be our guide, helping us out, regaining focus in our research and education, organized around claims about (anti-)determinism, (anti-)essentialism and (anti-)foundationalism. His method, in all its proximity to that of Hobbes, retains its commitment to freedom as a meaningful category, re-organized methodologically around the many precarious relations to nature that were and are involved.

Certainly, this is not the most practical goal and it is much easier to only claim the part that is about how the tragedy is deepening or how other people's elevated positions need to be brought low. Sometimes they do, and if this refers to Rousseau there are plenty of starting points. Often any type of commitment to a general will is out of reach, impractical in terms of everyday practices. He has no solution on offer that guarantees your individual freedom through some procedure, territorial claim or a concrete form of government. What he does, however, is present a compelling case for the need for freedom, exactly because the concept ends up, in a lot of settings, appearing as tragically insufficient at the level of the nations of millions, the metropolis, the geographical expanse of the natural world and so on.

Freedom, with Rousseau, will start small, and is only feasible in a few instances. Approached as a method, however, his formula shows a way out of the descend into relativism, localism and naive naturalism of contemporary STS and many of the forms of social science that

have followed its lead for a while now. We can, provisionally, acknowledge that there are tragic circumstances, that even though they are
empirical there is an inescapable need for theory, concepts and a
method that explains the approach and its politics. There's always a
political agenda, so let's at least not keep it implicit, apologetic and
as if it should be secondary. If it's part of the approach, it is not a
contradiction. It can be perfectly consistent to resist societal influence
on some type of science or technoscientific domain. It is, after all,
just another type of naturalism to be examined and the same applies
to stronger claims that assert that everything is social or to how we
should let nature back in, as Latour did.

When these types of naturalistic monism, as imperatives, are
removed, we can examine the contradictions that characterize the
world around us, the predicament that we are in. For example, we
might at once recognize that a certain scientific project has authority
and exercises it, while at the same time standing up for science when it
is under pressure. These are no longer incompatible when considered
in terms of how each invokes specific political theories of authority and
power. The solution is, therefore, a type of methodological pluralism
and on that condition, there is much that stays the same. There can be
restrained empiricism, bioconstitutionalism/s and "Critique". Each as
a means to an end. Even the original goal of STS is still there: to simply
show science and technology as part of social reality. So are the newer
goals, like advocating in favor of an openness to experimentation, as a
social value and as a commitment to an open future. At the same time,
however, there is a different type of interdisciplinarity that comes into
focus: one that has "strong" disciplinary components. Political theory
in its disciplinary form is only one example, it is a literature that has
no special status within the context of methodological pluralism. Its
presence as a discipline does not imply that there can be no sustained
interest in "inter", "multi" "trans", as each share the same suffix that
can be taken more or less seriously. Nevertheless, a "strong" presence
of political theory implies that STS, and adjacent interdisciplinary
fields, cannot hide their politics behind their preferred type of empiricism, like the imperative of "the case", laying down its claims while
suspending its reality.

This causes discomfort, there will be disagreements, a mismatch
of ontological commitments, that need to be worked through, as
scholarship, as political work, within the demos. STS and adjacent
types of social science will have to learn (again) to live without its
own consensus, because the alternative is worse, tragic really: an

infinite regress into ever more detailed claims and efforts to establish issue ownership, having to ever more opportunistically move along with the roles and content assigned in science policy, by funders and editorial boards. Rousseau would likely not be surprised to see particular interests being turned into conventions that rule academic fiefdoms, mandated by divided corporate and territorial sovereignties and with a hold over whatever social contracts there are. After all, his observations on the tragedy of his day did not include the coming Revolution. Similarly, any sense of urgency will likely seem overstated; after all, the masses, the majority and the elite of the community remain mostly oblivious to the urgency of today's tragedy, the one that surrounds us, that we are complicit in. If the tragedy is in full swing, and it's not unlikely, we should be realizing that, inevitably, our own legitimacy will be called into question. And, with Rousseau, it is giving meaning to freedom that carries this legitimacy in a crisis, and when at last we leave behind the protection of our empirical observations we might suddenly realize there is meaning, after all, and right at that time we, logically, find ourselves inept: "constitutively incapable of deciding" about anything (Tamminen and Deibel 2019, 67–70, see Agamben 2005).

"Upgrade in progress – don't shut down": might be the message that best captures our condition: the naive (naturalistic) acceptance of any and all updates to our many devices and the anxiety-ridden relation to the societal chains that bind us. We've been warned, over and over, that it is a tragedy, by McLuhan, Foucault, going back to the origins and now most of critical thinking is characterized by it. Let's take the little time there is and merge these warnings with a re-reading of Rousseau, re-examining the 21st-century return of such naivety, the return of "species of naturalism" that we continued to agree to blindly, even though we realize that restrictions are being imposed and we are being hooked up to the commercialization of everything. Let's return to Rousseau, his method and his commitment to freedom, which shows us a different way of how to (de-) hybridize our constitutional claims, our relation to truth, security, nature and democracy.

The result, for now, is that we are here no longer bound to consider our freedom in terms of the type of social contracting that are intrinsically tied to the state, the market or any other type of Hobbesian version of the "problem of sovereignty" that characterize our present (see Tamminen and Deibel 2019). It's our start, the one we got, and we need one. The type of hybrids we aim for, pedagogically, should not simply affirm that nature-cultures are many, that they mix and that this is the only reality. Their function is to re-orient our critical thinking (our Critique), in this case towards a conception of freedom in its relation to the Rousseauian dependency on things, whether code, seeds, or a more abstract relation to nature and technology. We have some time before the update that was promised is done, which is a frustrating moment because we know that it is not going to deliver, either security or truth, let alone freedom. Yet, when the moment will arrive that it delivers, the light will pop on again, and we might use that time to be in some type of position from where we can frantically start escaping from the regressive loop, we have gotten ourselves into; the one wherein things can be patched but not fixed, and where the (social) codes have gotten comfortably and conveniently out of control.

Notes

1 Arendt's overall argument is not only about the scientific outlook on the general will and its extension to the state and the market. At the start she identifies Rousseau's individualistic fiction of solitude with the decline of the status of "contemplation" from antiquity to the modern era, in the end we have her reflections on the meaning of freedom. From the perspective of this work, it is, therefore, a fully Rousseauian methodology and remains exemplary as political theoretical scholarship that deals with science and technology. This also applies to her discussion of "the household". This is a theme that opens the "discourse on political economy" (see Rousseau 2003, 128), in the sentence: "the meaning of the term was then extended to that great family, the state". In turn, it forms the theoretical core of Arendt's famous discussion of the public sphere, as indicated by the title of one of the sections: "the polis and the household" (see Arendt 1997, 28). See also Brooke (2012) for a specialist discussion on how stoic thought became part of political theory (and how not).

2 Constant said about Rousseau (as Arendt echoes before criticizing Marx): "certainly, I shall avoid the company of detractors of a great man. If I happen to agree with them on a single point I grow suspicious of myself; and in order to console myself for having seemed to be of their opinion… I feel I must disavow and keep these false friends away from me as much as I can" (Arendt 1997, 79).

3 The passage that follows is both famous and complex. It holds all the elements of this discussion in its density. It states: "The right of the first occupier, though more real than the right of the strongest, becomes a real right only when the right of property has already been established. Every man has naturally a right to everything he needs; but the positive act which makes him proprietor of one thing excludes him from everything else. Having his share, he ought to keep to it, and can have

no further right against the community. This is why the right of the first occupier, which in the state of nature is so weak, claims the respect of every man in civil society. In this right we are respecting not so much what belongs to another as what does not belong to ourselves" (Rousseau 2003a, 197).

4 Steadily this movement has been growing over the last 15 years. Starting as a discussion topic that only a few people, some of which academics, were interested in, surrounded by skeptic specialists. Soon after the specialists acknowledged its realism after the first initiative was launched, but, of course, it would never work. That also applied to the second, third and fourth example. The topic spreads gradually, and most recently many of these initiatives turned into a global coalition, see www.opensourceseeds.org/en/gossi (last checked February 2022). It might or might not have the impact it aims for, but academically speaking it has nicely demonstrated how fruitful it can be to be interested in freedom in whatever field one happens to be expected to study (see Deibel 2006).

References

Agamben, G. (2005). State of Exception. Chicago, University of Chicago Press.

Althusser, L. and Brewster B. (2007). Politics and History. Montesquieu, Rousseau, Marx. London, Verso.

Althusser, L. (1988). The social contract (the discrepancies). In Bloom, H. (Ed.). Jean Jacques Rousseau. New York, Chelsea House Publishers.

Arendt, H. (1997). The Human Condition. Chicago, University of Chicago Press.

Azuma, H. (2014). General Will 2.0: Rousseau, Freud, Google. New York, Vertical.

Baudrillard, J. (2000). The Vital Illusion. New York, Colombia University Press.

Benkler, Y. (2011). The Penguin and the Leviathan: the Triumph of Cooperation Over Self-Interest. New York, Crown Business books.

Berlin, I. (2002). Liberty. Oxford, Oxford University Press.

Bernardi, B. and Bensaude-Vincent, B. (2012). The presence of sciences in Rousseau's trajectory and works. In Grace, E. and Kelly, C. (Eds.). The Challenge of Rousseau. Cambridge, Cambridge University Press.

Berry, D.M. (2008). Copy, Rip, Burn: The Politics of Copyleft and Open Source. London, Pluto Press.

Birch, K. and Muniesa, F. (2020). Assetization: Turning Things into Assets in Technoscientific Capitalism. Cambridge: The MIT Press.

Bloom, H. (1988). Jean Jacques Rousseau. New York, Chelsea House Publishers.

Bogers, M., Afuah, A. and Bastian, B. (2010). Users as Innovators: A Review, Critique, and Future Research Directions. *Journal of Management, 36*(4), 857–875.

Bowker, G.C. (2005). Memory Practices in the Sciences. Cambridge MA, MIT Press.

Boyle, J. (2003). The second enclosure movement. *Law and Contemporary Problems, 66*(1), 33–74.

Brooke, C. (2012). Philosophical Pride: Stoicism and Political Though from Lipsius to Rousseau. Princeton, Princeton University Press.

Bull, H. (2002). The Anarchical Society: A Study of Order in World Politics. New York, Columbia University Press.

Burke, E. (2009). Reflections on the Revolution in France. Oxford, Oxford University Press.

Capra, F. and Mattei, U. (2015). The Ecology of Law: Toward a Legal System in Tune With Nature and Community. Oakland, Berrett-Koehler Publishers.

Collins, H (2004). Gravity's Shadow: the Search for Gravitational Waves. Chicago, Chicago University Press.

Cassirer, E. (1988). The question of Jean-Jacques Rousseau. In Bloom, H. (Ed.). Jean Jacques Rousseau. New York, Chelsea House Publishers.

Cassirer E. (1989). The Question of Jean Jacques Rousseau. New Haven, Yale University Press.

Collin. F. (2011). Science Studies as Naturalized Philosophy. Dordrecht, Springer.

Daston, L. and Park, K. (2001). Wonders and the Order of Nature, 1150–1750. New York, Zone Books.

Daston, L. and Elunbeck, E. (2011). Histories of Scientific Observation. Chicago, University of Chicago Press.

Dawkins, R. (1989). The Selfish Gene. Oxford, Oxford University Press.

Deibel, E. (2006). Common genomes: Open source in biotechnology and the return of common property. *Tailoring Biotechnologies*, *2*(2), 49–84.

Deibel, E. (2009). Common Genomes: On Open Source in Biology and Critical Theory Beyond the Patent. Amsterdam, VU University. Available at https://research.vu.nl/ws/portalfiles/portal/75951915/cover.pdf (checked 28 May 2022).

Deibel, E and Kloppenburg. J. (2015). Open licensing and plant varieties: On creating a protected commons. In Thomas, F., Bonneuil, C. and Boisvert, V. (Eds.). Le Pouvoir De La Biodiversité. Paris, Editions Quae Paris.

Deibel, E. (2013). Open variety rights: Rethinking the commodification of plants. *Journal of Agrarian Change*, *13*(2), 282–309.

Deibel, E. (2014). Open genetic code: on open source in the life sciences. *Life Sciences, Society and Policy*, *10*(2), 1–23.

Deleuze, G. (1987). A Thousand Plateaus: Capitalism and Schizophrenia. Minneapolis, University of Minnesota Press.

Delfanti, A. (2013). Biohackers: The Politics of Open Science. London, Pluto Press.

Dent, N. (2005). Rousseau. New York, Routledge.

Derrida, J. (1976). Of Grammatology. Baltimore, Johns Hopkins University Press.

Derrida, J. (1988). ....That dangerous supplement.... In Bloom, H. (Ed.). Jean Jacques Rousseau. New York, Chelsea House Publishers.

Dragona D. and Charitos D. (2016). Going off-the-cloud: The role of art in the development of a user-owned & controlled connected world. *The Journal of Peer Production*, *9*. http://peerproduction.net/issues/issue-9-alternative-internets/peer-reviewed-papers/going-off-the-cloud/

Dupre J. (2005). Darwin's Legacy: What Evolution Means Today. Oxford, Oxford University Press.

Fontana, B. (1991). Benjamin Constant and the Post-Revolutionary Mind. New Haven, Yale University Press.

Feenberg, A. (1999). Questioning Technology. London, Routledge.

Feyerabend, P. (2010). Against Method. London, Verso.

Foucault, M. (2002). The Order of Things: An Archaeology of the Human Sciences. London, Routledge.

Foucault, M. (2003). Society Must Be Defended. New York, Picador.

Frickel S. (2006). The New Political Sociology of Science: Institutions, Networks, and Power. In Frickel S. and Moore K. (Eds.). Madison, University of Wisconsin Press.

Fukuyama, F. (2002). Our Posthuman Future: Consequences of the Biotechnology Revolution. New York, Picador.

Gieryn, T.F. (1983). Boundary-work and the demarcation of science from non-science: Strains and interests in professional ideologies of scientists. *American Sociological Review, 48*(6): 781–795.

Gozzi, G. and Valente, F. (2021). Humanitarian Intervention, Colonialism, Islam, and Democracy: an Analysis Through the Human-Nonhuman Distinction. London, Routledge.

Habermas, J. (2003). The Future of Human Nature. Cambridge, Polity.

Halffman W. (2003). Boundaries of regulatory science: Eco/toxicology and aquatic hazards of chemicals in the US, England and the Netherlands. Amsterdam (Dissertation), University of Amsterdam.

Harway D.J. (1991). Simians, Cyborgs, and Women: The Reinvention of Nature. New York, Routledge

Haraway, D.J. (1997). Modest_Witness@Second_Millennium.FemaleMan©_Meets_Oncomouse™: Feminism and Technoscience. New York, Routledge.

Hardt, M. and Negri, A. (2000). Empire. Cambridge, Harvard University Press.

Harvey, D. (1996). Justice, Nature & the Geography of Difference. Oxford, Blackwell Publishers.

Harvey, D. (2011). The future of the commons. *Radical History Review, 109*, 101–110.

Hayles, N.K. (1999). How We Became Posthuman: Virtual Bodies in Cybernetics, Literature, and Informatics. Chicago, The University of Chicago Press.

Higgins, E. (2021). We Are Bellingcat: An Intelligence Agency for the People. London, Bloomsbury.

Hobbes, T. (1985). Leviathan – Edited With an Introduction by C.B. Macpherson. London, Penguin Books.

Hoffman, S. and Fidler, D.P (1991). Rousseau on International Relations. Oxford, Oxford Clarendon Press.

Horkheimer, M. (2003 [1944]). Eclipse of Reason. New York, NY, Continuum.

Horkheimer, M. and Adorno, T.W. (2002). Dialectic of Enlightenment. Stanford, CA, Stanford University Press.

Howard Campbell, S. and Scott, John T. (2005). Rousseau's politic argument in the discourse on the sciences and arts. *American Journal of Political Science, 49*(4), 818–828.

Hume, D. (2011). A Treatise of Human Nature. Volume I. Oxford, Oxford University Press.

Hunter, M.C.W. (Ed.) (1994). Robert Boyle Reconsidered. Cambridge, Cambridge University Press.

Huntington, S. (2002). The Clash of Civilization and the Remaking of World Order. London, The Free Press.

Jasanoff, S. (Ed.) (2011). Reframing Rights: Bioconstitutionalism in the Genetic Age. London, MIT Press.

Jasanoff, S. (2016). The floating ampersand: STS past and STS to come. *Engaging Science, Technology, and Society*, *2*(1), 227–237.

Kant, I., Humphrey, T. and Kant, I. (1983). Perpetual Peace, and Other Essays on Politics, History, and Morals. Cambridge, Hacket Publishing Company.

Kaplan, R.D. (2002). The Coming Anarchy: Shattering the Dreams of the Post-Cold War. New York, Vintage Books.

Kelly, C. (2012). Rousseau and the illustrious Montesquieu. In Grace E. and Kelly C. (Eds.). The Challenge of Rousseau. Cambridge, Cambridge University Press.

Kelty, C.M. (2008). Two Bits: The Cultural Significance of Free Software. Durham, Duke University Press.

Kloppenburg, J. (2005). First the Seed: The Political Economy of Plant Biotechnology. Madison, University of Wisconsin Press.

Kuhn, T.S. (2012). The Structure of Scientific Revolutions. Chicago, University of Chicago Press.

Kymlicka, W. (2002). Contemporary Political Philosophy: An Introduction. Oxford, Oxford University Press.

Latour, B. (1993). We Have Never Been Modern. New York, Harvester Wheatsheaf.

Latour, B. (1999). Pandora's Hope: Essays on the Reality of Science Studies. Cambridge, Harvard University Press.

Latour, B. (2017). Facing Gaia: Eight Lectures on the New Climatic Regime. Cambridge, Polity Press.

Law. J. (2008). On sociology and STS. *The Sociological Review*, *56*(4), 623–649.

Law J. (2015). STS as method. In Felt, U., Fouché, R., Miller, K. and Smith-Doerr, L. The Handbook of Science and Technology Studies. Cambridge, MA: The MIT Press.

Lelja, P (2018). Defending a common world: Hannah Arendt on the state, the nation and political education. *Studies in Philosophy and Education, 37*, 537–552.

Levi-Strauss, C. (1961). Tristes Tropiques. New York, Criterion Books.

Lindsay P. (2016). Thinking back (and forward) to Rousseau's Emile. *Journal of Political Science Education*, *12*(4), 487–497.

Luhrmann, T.M. (1990). Our master, our brother: Lévi-Strauss's debt to Rousseau. *Cultural Anthropology*, *5*(4), 396–413.

MacPherson, C.B. (1962). The Political Theory of Possessive Individualism: Hobbes to Locke. Oxford, Oxford University Press.

Maistre, de J.M. and Lebrun, R. (1996). Against Rousseau: "On the State of Nature"; and "On the Sovereignty of the People". Montreal, McGill-Queen's University Press.

Marcuse, H. (1991). One-dimensional Man: Studies in the Ideology of Advanced Industrial Society. London, Routledge.

Marx, K. and Engels, F. (1969). Manifesto of the communist party. In Marx/ Engels Selected Works, Volume 1, Moscow: Progress Publishers. https://www.marxists.org/archive/marx/works/1848/communist-manifesto/index.htm

Marx, K. (1990). Capital: A Critique of Political Economy. Volume 1. London, Penguin Group.

May, C. (2000). The Global Political Economy of Intellectual Property Rights: The New Enclosures? London, Routledge.

Medina, E. (2011). Cybernetic Revolutionaries: Technology and Politics in Allende's Chile. Cambridge, The MIT Press.

McLuhan, M. (1995). Understanding Media: The Extensions of Man. London: Routledge.

Merton, R.W. (2017). The Sociology of Science: Theoretical and Empirical Investigations. Chicago, University of Chicago Press.

Moglen, E. (2003). Anarchy triumphant: Free software and the death of copyright. *First Monday*, *4*(8). https://firstmonday.org/ojs/index.php/fm/article/view/684

Muthu, S. (2003). Enlightenment Against Empire. Princeton, Princeton University Press.

Nahuis, R. and van Lente H. (2008). Where are the politics? Perspectives on democracy and technology. *Science, Technology & Human Values*, *33*(5), 559–581.

Hammer, C. (2015). Goethe and Rousseau: Resonances of the Mind. Lexington, The University Press of Kentucky.

Hurlbut, J.B., Jasanoff, S. and Saha, K. (2020). Constitutionalism at the nexus of life and law. *Science, Technology, & Human Values*, *45*(6), 979–1000.

Neuhauser, F. (2003). Foundations of Hegel's Social Theory: Actualizing Freedom. New Haven, Harvard University Press.

Neuhauser, F. (2014). Rousseau's Critique of Inequality. Cambridge, Cambridge University Press.

Nietzsche F. (2003). Beyond Good and Evil Paperback. London, Penguin.

Ohana D. (2017). Jean-Jacques Rousseau and the Promethean Chains. *Politics, Religion & Ideology*, *18*(4), 383–408.

Ostrom, E. (1990). Governing the Commons: the Evolution of Institutions for Collective Action. Cambridge, University Press.

Pagden, A. (2013). The Enlightenment and Why It Still Matters. New York, Random house.

Parry, B. (2004). Trading the Genome: Investigating the Commodification of Bio-Information. New York, Colombia University Press.

Petrovic, J.E. and Rolstad K. (2016). Educating for autonomy: Reading Rousseau and Freire toward a philosophy of unschooling. *Policy Futures in Education*, *15*(7): 817–833.

Popper, K. (2002). The Open Society and Its Enemies. London, Routledge.

Rawls, J. (1999). A Theory of Justice. Oxford, Oxford University Press.

Raymond, E. (1999). The Cathedral and the Bazaar: Musings on Linux and Open Source by an Accidental Revolutionary. Sebastopol, O'Reilly.

Roulston, C, (1989). Virtue, Gender, and the Authentic Self in Eighteenth-Century Fiction. Gainesville, University Press of Florida.

Roosevelt, G.G. (1990). Reading Rousseau in the Nuclear Age. Philadelphia, Temple University Press.

Rosenfeld, D. (1987). Rousseau's unanimous contract and the doctrine of popular sovereignty. *History of Political Thought*, *8*(1), 83–110.

Rousseau, J.J. (1993). Émile. London, Everyman. Reprint 1992, original edition published 1911.

Rousseau, J.J. (1997). Julie, or, the New Heloise Letters of Two Lovers Who Live in a Small Town at the Foot of the Alps. Hanover, University Press of New England.

Rousseau, J.J. (2003). The Social Contract and Discourses. London, Everyman. Reprint 1993, original edition published 1913.

Santo L.A. (1994). Rousseau's Art of Persuasion in La Nouvelle Héloïse. Lanham, University Press of America.

Serres, M. (1995). The Natural Contract. Ann Arbor, University of Michigan Press.

Schiller, Friedrich (2016). On the Aesthetic Education of Man. London, Penguin Classics.

Scot, T. (2006). Jean Jacques Rousseau: Critical Assessments of Leading Political Philosophers. New York, Routledge.

Sell, S. (2007). Intellectual property and the Doha development round. In Lee, D. and Wilkinson, R. (Eds.). The WTO After Hong Kong: Progress in, and Prospects for, the Doha Development Agenda. New York, Routledge.

Shapin, S. and Schaffer, S. (1985). Leviathan and the Air-Pump: Hobbes, Boyle, and the Experimental Life. Princeton, Princeton University Press.

Shell, S. and Vekley, R. (2017). Rousseau and Kant: Rousseau's Kantian legacy. In Rosenblatt, H. and Schweigert, P. (Eds.). Thinking with Rousseau. Cambridge, Cambridge University Press.

Simpson, M. (2006). Rousseau's Theory of Freedom. London, Continuum.

Skinner, Q. (1996). Reason and Rhetoric in the Philosophy of Hobbes. Cambridge, Cambridge University Press.

Söderberg, J. (2017). The genealogy of "empirical post-structuralist" STS, retold in two conjunctures: The legacy of Hegel and Althusser. *Science as Culture*, *26*(2): 185–208.

Stephenson, N. (1999). In the Beginning... Was the Command Line. New York, Avon Books.

Stilgroe, J., Owen, R. and Macnaghten, P. (2013). Developing a framework for responsible innovation. *Research Policy*, *42*(9), 1568–1580.

Sunder, M. (2007). The invention of traditional knowledge. *Law and Contemporary Problems*, *70*(9), 97–12.

Sunder Rajan, K. (2006). Biocapital: The Constitution of Postgenomic Life. Durham, Duke University Press.

Tamminen, S. and Brown, N. (2011). Nativitas: Capitalising genetic nation-hood. *New Genetics and Society*, *30*, 73–99.

Tamminen and Deibel (2019). Recoding Life: Information and the Biopolitical. New York, Routledge.

Tully, J. (1995). Strange Multiplicity: Constitutionalism in an Age of Diversity. Cambridge, Cambridge University Press.

Tuck, R. (1999). The Rights of War and Peace. Oxford, Oxford University Press.

Tuck, R. (2017). Rousseau and Hobbes. The Hobbesianism of Rousseau. In Rosenblatt, H. and Schweigert, P. (Eds.). Thinking With Rousseau. Cambridge, Cambridge University Press.

Ucaryilmaz Deibel, T. (2022). Corporate social responsibility in the legal frame-work of global value chains. *Law and Development Review*, *15*(2), 329–356.

Vermeulen, N., Tamminen, S. and Webster, A. (2012). Bio-Objects: Life in the 21st Century. Farnham, Ashgate.

Virilio, P. and Lotringer, S. (1983). Pure War. Los Angeles, Semiotext(e).

von Hippel, E. (1988). The Sources of Innovation. New York, Oxford University Press.

von Hippel, E. (Ed.) (2005). Democratizing Innovation. Cambridge, MIT Press.

Wain, K. (2014). Between Truth and Freedom: Rousseau and Our Contemporary Political and Educational Culture. London, Routledge.

Weiss, A (1987). Rousseau, antifeminism, and woman's nature. *Political Theory*, *15*(1), 81–98.

Wendt, A. (1999). Social Theory of International Politics. Cambridge, Cambridge University Press.

Williams, M.C. (2005). The Realist Tradition and the Limits of International Relations. Cambridge, Cambridge University Press.

Winner, L. (1982). Techne and Politeia: The Technical Constitution of Society. Dordrecht, D. Reidel Publishing Company.

Wright, J.K. (2017). Rousseau and Montesquieu. In Rosenblatt, H. and Schweigert, P. (Eds.). Thinking with Rousseau. Cambridge, Cambridge University Press.

Wynn, B. (2008). Public participation in science and technology: Performing and obscuring a political-conceptual category mistake. *East Asian Science, Technology and Society: An International Journal*, *1*(1), 99–112.

Index